Maurice Thompson

Byways and Bird Notes

Maurice Thompson

Byways and Bird Notes

ISBN/EAN: 9783744652322

Printed in Europe, USA, Canada, Australia, Japan

Cover: Foto ©berggeist007 / pixelio.de

More available books at **www.hansebooks.com**

BY-WAYS

AND

BIRD NOTES

BY

MAURICE THOMPSON

AUTHOR OF

"At Love's Extremes," "His Second Campaign," "Songs
of Fair Weather," "A Tallahassee
Girl," Etc.

NEW YORK
JOHN B. ALDEN, PUBLISHER
1885

TROW'S
PRINTING AND BOOKBINDING COMPANY,
NEW YORK.

CONTENTS.

BY-WAYS

AND

BIRD-NOTES.

IN THE HAUNTS OF THE MOCKING-BIRD.

THE mocking-bird has been called the American nightingale, with a view, no doubt, to inflicting a compliment involving the operation, known to us all, of damning with faint praise. The nightingale presumably is not the sufferer by the comparison, since she holds immemorial title to preëminence amongst singing-birds. The story of Philomela, however, as first told, was not an especially pleasing one, and the poets made no great use of it. Nowhere in Greek or Roman literature, so far as I know, is there any genuine lyric apostrophe to the nightingale comparable to Sappho's fragment *To the Rose;* still the bird has a prestige gathered from centuries of poetry and upheld by the master romancers of the world.

To compare the song of any other bird with that of the nightingale is like instituting a comparison between some poet of to-day and Shakespeare, so far as any sympathy with the would-be rival is concerned. The world has long ago made up its mind, and when the world once does that there is an end, a *cul de sac*, a stopping-place, of all argument of the question. Indeed, it is a very romantic distance that separates the bird from most of us. Chaucer's groves and Shakespeare's woods shake out from their leaves a fragrance that reaches us along with a song which is half the bird's and half the poet's. We connect the nightingale's music with a dream of chivalry, troubadours, and mediæval castles. It is as dear to him who has heard it only in the changes rung by the Persian, French, and English bards as it is to him whose chamber window opens on a choice haunt of the bird in rural England.

I might dare to go further and claim that I, who have never heard a nightingale sing, can say with truth that its music is, in a certain way, as familiar to me as the sound of a running stream or the sough of a spring breeze. I often find myself reluctantly shaking off something like a recollection of having somewhere, in some dim old grove, heard the voice that Keats imprisoned in his matchless ode. There is a sort of aerial perspective in the mere name of the nightingale; it is like some of those classical allusions which bring into a modern essay suggestions with an infinite distance in them. So thoroughly has this been felt that it may safely be said that the nightingale has been more frequently mentioned by our American writers, good, bad, and indifferent, than any one of our native birds. No doubt it ought to provoke a smile, this gushing about a music one has never heard; but, like the music of the spheres and the roar of the ocean, the nightingale's voice is common property, and we all take it as a sort of hered-

itary music, descending to us by immemorial
custom. Its notes are echoing within us, and
we feel their authenticity though in fact we
know as little about the bird as chemists do
about Geber. How shall we doubt that the
bird whose song inspired Keats to write that
masterpiece of English poetry is indeed a
wonderful musician? Shakespeare and rare
Ben Jonson and Burns and Scott and Shelley
and Byron heard this same song; it was just
as clear and sweet as it is now when Chaucer
was telling his rhymed tales, when Robin
Hood was in the greenwood, even when the
Romans made their first invasion.

In a general way, we do not think of the
nightingale having a nest and rearing a brood
and dying. It is simply the incomparable
nightingale, philomela, rossignol, or whatever
the name may be,—a bird that has been singing in rose-gardens and orange-orchards and
English woods night after night for thousands
of years without a rival. Its song is to the
imagination of all of us

"L'hymne flottant des nuits d'été."

as Lamartine has expressed it. So it can
easily be understood how hard a struggle our
American mocking-bird is going to have before
it reaches a place in the world's esteem beside
the nightingale. Nor is it my purpose to do
anything with a special view to aid it in the
struggle; but I have studied our bird in all
its haunts and in all seasons, with a view to a
most intimate acquaintance with its habits, its
song, and its character.

To begin with, the name *mocking-bird*, is a
heavy load for any bird to bear. Unmusical

as it is, the worst feature of such an appella-
tion is the idea of flippancy and ill-breeding
that it conveys. To "mock" is to imitate
with an ill-natured purpose, to jeer at, to ridi-
cule; it was for mocking that bad children
were made food for bears. Such a name
carries with it a shadow of something repellant, and no poet can ever rescue it, as a
name, from its meaning and its eight harsh
consonants. It would indeed require some
centuries of romantic and charming associa-
tions to make of it a name by which to conjure, as in the case of the nightingale. The
bird, with almost any other name than mocking-bird, would fare much better at the hands
of artists and poets, and might hope, if birds
may hope at all, finally to gain the meed of
praise it so richly deserves.

In a beautiful little valley among the mountains of North Georgia I first began to study
the mocking-bird in its wild state. It was not
a very common bird there, just rare enough to
keep one keenly interested in its habits. I
had great trouble in finding a nest. Many a
delightful tramp through the thorny thickets
and wild orchards of plum-trees ended in nothing, before my eyes discovered the loose sticks
and matted midribs of leaves which usually
make up the songster's home. The haw-tree,
several varieties of which grow in the glades
of what is known as the Cherokee Region, is a
favorite nesting-place, and so is the honey-
locust tree, which is also much chosen by the
shrike or butcher-bird. There is so strong a
resemblance in colors and size between this
shrike and the mocking-bird that one is often
mistaken for the other by careless observers,

hence in some neighborhoods, I have found a
strong prejudice existing against the mocking-
bird on account of the fiendish habits of the
shrike.

A mountain lad once led me over a considerable mountain and down into a wild dell
to show me a nest in a thorn tree, where he
was sure I should find every evidence that a
mocking-bird was a soulless monster, murder-
ing little pee-wee fly-catchers and warblers,
and impaling them on thorns out of sheer
wantonness. I felt sure it was a shrike, but
the boy said he knew better. Didn't he know
a mocking-bird when he saw it? He had
heard it sing and "mock" all the birds in the
thickets around, and had also seen it doing its
brutal work. Boys are sometimes very close
and reliable in their observations, and this one
was an inveterate hunter, and so stoutly as-
serted his knowledge that I was induced to
test his accuracy by going with him to the
place he called Mocking-Bird Hollow. Of
course the nest was that of a shrike, but a
number of mocking-birds were breeding in the
immediate vicinity, hence the mistake.

The mocking-bird does not appear to be a
strictly migratory bird, its range being much
narrower than that of the brown-thrush, the
cat-bird, and the wood-thrush. I have never
been able to find it a regular visitant in the
West north of Tennessee, though I have no
reason to doubt that it comes at times much
farther, even into the Ohio valley. In the
mountain valleys it is extremely wary and shy,
its habits approaching very close to those
attributed to the nightingale of England. It
chooses lonely and almost inaccessible nest-

ing places, and will not sing if at all disturbed.
Often, while I have been lying on the ground
in some secluded glade, I have heard, far in
the night, a sudden gush of melody begun by
one bird and echoed by another and another
all around me, filling the balmy air of spring
with a half-cheerful, half-plaintive medley.
This is more common when the moon shines,
but I have heard it when the night was black.

At several points near the coast of the
Carolinas I have found the mocking-bird apparently a resident, and yet, so far South as
Savannah, Georgia, it seems to shrink from the
occasional midwinter rigors. In the hills near
the Alabama River, not far from Montgomery,
it is certainly resident, but I found it a much
shyer bird there than in the thickets along the
bayous of Louisiana. Early in the winter of
1883 I made a most careful search for the
mocking-bird in Pensacola, Florida, and its
environs, but found none. I was told that the
bird would appear about the last of February.
At Marianna, Florida, and along the line of
the road thence to the Appalachicola River, I
saw it frequently in midwinter. On the Gulf
Coast, down as far as Punta Rassa, and across
the peninsula to the Indian River country, in
the orange, lemon, and citron groves, in the
bay thickets, and even in the sandy pine
woods, I noted it quite frequently. In this
semi-tropical country it is not so shy and so
chary of its song, as it is farther north. Near
the mouth of the St. Mark's River, as I lay under a small tree, a mocking-bird came and lit
on the top of a neighboring bush, and sang for
me its rarest and most wonderful combination,
called by the negroes the "dropping song."

Whoever has closely observed the bird has noted its "mounting song," a very frequent performance, wherein the songster begins on the lowest branch of a tree and appears literally to mount on its music, from bough to bough, until the highest spray of the top is reached, where it will sit for many minutes flinging upon the air an ecstatic stream of almost infinitely varied vocalization. But he who has never heard the "dropping song" has not discovered the last possibility of the mocking-bird's voice. I have never found any note of this extremely interesting habit of the bird by any ornithologist, a habit which is, I suspect, occasional, and connected with the most tender part of the mating season. It is, in a measure, the reverse of the "mounting song," beginning where the latter leaves off. I have heard it but four times, when I was sure of it, during all my rambles and patient observations in the chosen haunts of the bird; once in North Georgia, twice in the immediate vicinity of Tallahassee, Florida, and once near the St. Mark's River, as above mentioned. I have at several other times heard the song, as I thought, but not being able to see the bird, or clearly distinguish the peculiar notes, I cannot register these as certainly correct. My attention was first called to this interesting performance by an aged negro man, who, being with me on an egg-hunting expedition, cried out one morning, as a burst of strangely rhapsodic music rang from a haw thicket near our extemporized camp, "Lis'n, mars, lis'n, dar, he's a droppin', he's a-droppin', sho's yo' bo'n!" I could not see the

bird, and before I could get my attention
rightly fixed upon the song it had ended.

Something of the rare aroma, so to speak,
of the curiously modulated trills and quavers
lingered in my memory, however, along with
Uncle Jo's graphic description of the bird's
actions. After that I was on the lookout for
an opportunity to verify the negro's state-
ments.

I have not exactly kept the date of my first
actual observation, but it was late in April, or
very early in May; for the crab-apple trees,
growing wild in the Georgian hills, were in
full bloom, and spring had come to stay. I
had been out since the first sparkle of day-
light. The sun was rising, and I had been
standing quite still for some minutes, watching a mocking-bird that was singing in a
snatchy, broken way, as it fluttered about in a
thick-topped crab-apple tree thirty yards distant from me. Suddenly the bird, a fine specimen, leaped like a flash to the highest spray
of the tree and began to flutter in a trembling,
peculiar way, with its wings half-spread and
its feathers puffed out. Almost immediately
there came a strange, gurgling series of notes,
liquid and sweet, that seemed to express utter
rapture. Then the bird dropped, with a backward motion, from the spray, and began to
fall slowly and somewhat spirally down through
the bloom-covered boughs. Its progress was
quite like that of a bird wounded to death by a
shot, clinging here and there to a twig, quivering and weakly striking with its wings as it fell,
but all the time it was pouring forth the most
exquisite gushes and trills of song, not at all
like its usual medley of improvised imitations

but strikingly, almost startlingly, individual and unique. The bird appeared to be dying of an ecstasy of musical inspiration. The lower it fell the louder and more rapturous became its voice, until the song ended on the ground in a burst of incomparable vocal power. It remained for a short time, after its song was ended, crouching where it had fallen, with its wings outspread, and quivering and panting as if utterly exhausted; then it leaped boldly into the air and flew away into an adjacent thicket.

Since then, as I have said, three other opportunities have been afforded me of witnessing this curiously pleasing exhibition of bird-acting. I can half imagine what another ode Keats might have written had his eyes seen and his ears heard that strange, fascinating, dramatically rendered song. Or it might better have suited Shelley's powers of expression. It is said that the grandest bursts of oratory are those which contain a strong trace of a reserve of power. This may be true; but is not the best song that wherein the voice sweeps, with the last expression of ecstasy, from wave to wave of music until with a supreme effort it wreaks its fullest power, thus ending in a victory over the final obstacle, as if with its utmost reach? Be this as it may, whoever may be fortunate enough to hear the mocking-bird's "dropping song," and at the same time see the bird's action, will at once have the idea of genius, pure and simple, suggested to him.

The high, beautiful country around Tallahassee, in Middle Florida, is the paradise of mocking-birds. I am surprised to find this

region so little visited, comparatively speaking,
by those who really desire to know all that is
beautiful and interesting in our country. Perhaps it is because the places most frequented
by the mocking-bird have not been sought by
those deeply interested in bird-habits and
history, that so little is known of the most
striking traits of its character. Quite certain
it is that no monograph exists which gives to
the general reader any approximate idea of
our great American singer. I must say just
here that the mocking-bird's song in captivity,
strong and sweet as it is, and its voice from
the cage, liquid, flexible, and pure, are not
in the least comparable to what they are in the
open-air freedom of a Southern grove. If you
would hear these at their best, and they are
truly worth going a long journey to hear, you
must seek some secluded grove in Southern
Alabama, Georgia, or Middle Florida about
the last of March or the first of April, when
spring is in its prime and the gulf breezes are
flowing over all that semi-tropical region.

It is a silly notion, without any foundation
in fact, that the mocking-bird in its wild state
is a mere mimic, without a song of its own.
The truth is that all birds get their notes, as
we get our language, by imitating what they
hear. Very few of them, however, are sufficiently gifted mentally and vocally to be able
to pass the limitation of immemorial heredity,
or to feel any impulse toward any attainments
of voice beyond what they catch as younglings
from their parents. Hence, as a rule, the
young bird is satisfied with the pipes and calls
caught from its immediate ancestors. No
doubt a lack of finely developed vocal organs

has much to do with this. But the mocking-bird, the brown-thrush, and the cat-bird are notable exceptions to the rule. Nature has endowed them with an instinctive impulse toward a cultivation of their vocal powers, as well as with voices capable of wonderful achievements.

A mocking-bird reared in captivity becomes much more a mere mimic than the wild bird, and yet, so strong is the hereditary tendency, the caged bird will perfectly sound the notes of a grossbeak or a blue-jay without ever having heard them. I have heard a mocking-bird, reared in a cage in Indiana, utter with singular accuracy the cry of the Southern wood-pecker (*Picus querulus*), a bird I have never seen north of the Cumberland Mountains.

Many little incidents noted in the woods and in the orchards haunted by the mocking-bird have led me to conclude that a genuine sense of the importance of singing well inspires some of its most remarkable efforts. One morning in March, 1881, I looked out of a window in the old City Hotel at Tallahassee, and witnessed a pitched battle of song between a brown-thrush and a mocking-bird. In the grounds about the Capitol building across the street stood some venerable oak trees just beginning to leave out. The birds had each chosen a perch on the highest practicable point of a tree. They were not more than fifty feet apart, and with swelling throats were evidently vying fiercely with each other. This gave me the best possible opportunity of comparing their styles and methods of expression. To my ear the brown-thrush in the wild state is a sweeter singer than any caged mocking-bird; but when both are free, the latter is infinitely superior at every point. There is a wide variety of pure flute-notes expressed by the wild mocking-bird. These notes become vitiated in captivity and their tone degraded to the level of mere mellow piping. In the hedges of Cherokee rose that grew along the old Augustine road east of Tallahassee, mocking-birds were so numerous that their songs, mingling together, made a strange din which could be heard a long way on a still morning.

I have already spoken of the injustice done the mocking-bird by the name given it, but at this point I may say that other American song birds of a superior order have suffered even more from this cause. *Cat-bird* and *thrasher*, —what names to be embalmed in poetry and romance! It required all the genius of Emerson successfully to use a titmouse as the subject for a poem. If Bryant's *Lines to a Waterfowl* had been addressed to a duck or a snakebird, one would scarcely be content to accept the poem as perfect. A name certainly has an intrinsic value.

Mr. Cable in his powerful novel, *Dr. Sevier*, speaks of the mocking-bird's morning note as unmusical. At certain seasons of the year the bird's voice is not especially pleasing, but this is not in song-time. Early morning and the twilight of evening in the spring call forth its most charming powers. Its night song is sweet and peculiarly effective, but except on rare occasions in the nesting season, when the moon is very brilliant the nocturnal notes are pitched in a minor key and the voice is less flexible and brilliant, as if the bird were singing in its sleep.

In Florida and in the valley of the Alabama,
I observed the mocking-bird assuming a familiarity with man very closely approaching voluntary domestication. A pair had their nest in
a small vine-covered peach-tree close to the
window of a room for some weeks occupied by
me. They seemed not in the least disturbed
when I boldly watched them, though occasionally the male bird was inclined to scold if I
raised the window. Every morning, just at
the peep of dawn, the singing began, and was
kept up at intervals all day. The house was
a mere cabin with unchinked cracks. All out-
door sounds came in freely. The Suwanee
River, made famous by the *Old Folks at
Home*, rippled near, and the heavy perfume of
magnolia flowers filled the air. My vigorous
exercise in the woods and fields by day, which
was sometimes continued far into the night,
made me sleep soundly, but very often I was
aroused sufficiently to be aware of a nocturne,
all the sweeter to my half-dreaming sense on
account of its plaintive and desultory rendering.

In the neighborhood of Thomasville, Georgia, a mocking-bird's nest, built in a pear-
tree, was close to a kitchen door, where ser-
vants were all day passing in and out within
ten or twelve feet of the sitting bird. The
brood was hatched, and the young taken by a
negro and sold to a New York tourist for
twenty dollars. The birds tore up their nest
as soon as it was robbed, and appeared greatly
excited for a few days; but one morning the
singing began again, and soon after a new
nest was built a little higher up in the same
tree.

2

It has been told of the mocking-birds that, in Louisiana and other Southern regions, when such of them as have taken a summer jaunt to New England or Pennsylvania return to the magnolia and orange groves in late autumn, they are attacked by their resident brethren. My observation has not tended to verify this. Nor can I bear testimony to the bravery and fighting qualities of the mocking-bird. The blue-bird whips it, driving it hither and yon at will, though not more than half its size. It is, however, a famous scold and blusterer, accomplishing a good deal by fierce threats and savage demonstrations. I do not believe the story about it killing snakes. It would be a very small and weak reptile that such a bird could kill, being so poorly armed for warlike exploits.

On a pedestrian tour through the loveliest and loneliest part of Middle Florida, I was struck with the strong contrast between the negroes and the white people as to the extent and accuracy of their ornithological knowledge, a contrast almost as marked as that of color. I could get no information from the whites. They had never paid any attention to mocking-birds. The subject appeared to them too slight and trivial to be worth any study. But the negroes were sometimes enthusiastic, always interested and interesting. Somehow there has always seemed to me a fine touch of power in the way a cabin, a few banana-stalks, a plum-tree or two, and a straggling bower of grape-vines get themselves together for the use of indolent negroes and luxury-loving mocking-birds. I have fancied it, or else there is a marked preference shown by the songster for the cots of the freedmen, and there can be no doubting that a warm feeling for the bird is nursed by the ordinary negro.

As I have suggested, the nature of the mocking-bird is that of a resident more than that of a migratory bird, and I am inclined to name its true habitat semi-tropical. Even so far South as Macon, Ga., and in the region of Montgomery, Ala., the chilly days of midwinter are sufficient to drive the birds to heavy cover. In fact, a large majority of the species of *Mimus* (*Mimus polyglottus* being the scientific name of the mocking-bird) are to be found in South America and in the tropical islands of the Atlantic.

The plantation negroes used to have a saying which might serve the turn of Mr. Harris or Mr. Macon : " Takes a red-hot sun fo' ter bri'l de mockin'-bird's tongue, but er mighty small fros' er gwine ter freeze 'im froat up solid." Mr. Fred. A. Ober, in his report of explorations made in the Okeechobee region, does not mention seeing the mocking-bird, but it is there, nevertheless, or was in 1867. I remember seeing a fine fellow flying about in some small bushes, near the remains of a deserted cabin, on the north-eastern shore of the lake. I saw some paroquets at the same place.

On what is known as the Dauphine Way, running west from Dauphine Street in Mobile, mocking-birds used to be numerous, nesting in the groves on either side and filling the air with their songs. Whoever has walked out on this lovely road will remember a low, old-fashioned brick house, no doubt a plantation residence one day, with a row of queer little dormer windows on the roof in front, and graduated parapets to hide the gables, a long lean-to veranda and a row of chimneys, a dark, heavy-looking building near the south side of the Way. In a small tree just east of this house used to sing a mocking-bird whose voice was as much above the average of his kind as Patti's voice is above the average woman's voice. If one could get a caged bird to sing as that one did, he might profitably advertise it for concerts. A friend and I sat down across the Way from the house, and, while the gulf breeze poured over us and the bird music filled our ears, got a sketch of the charmingly picturesque old place ; but somehow we could not put in the song of the wonderful mocking-bird.

Bird-fanciers and bird-buyers may profit by what I now whisper to them, to wit : the best-voiced mocking-birds, without a doubt, are those bred in Middle Florida and Southern Alabama. I have no theory in connection with this statement of a fact; but if I were going to risk the reputation of our country on the singing of a mocking-bird against a European nightingale, I should choose my champion from the hill-country in the neighborhood of Tallahassee, or from the environs of Mobile.

No doubt proper food has much to do with the development of the bird in all its parts, and it may be that the dry, fertile, chocolate-tinted hills that swell up along the Gulf Coast produce just the berries, insects, and other tid-bits needed for the mocking-bird's fullest growth. Then, perhaps, the climate best suits the bird's nature. Be this as it may, I have found no birds elsewhere to compare with those in that belt of country about thirty miles wide, stretching from Live Oak in Florida, by way of Tallahassee, to some miles west of Mobile. Nor is there anywhere a more interesting country to him who delights in pleasant wildwood rambles, unusual scenery, and a wonderful variety of birds and flowers in their season.

Most of our descriptive ornithologists have taken great pains to assure their readers that the American mocking-bird is very plain, if not positively unattractive in its plumage. But to my eye the graceful little fellow, especially when flying, is an object of real beauty. There is a silver-white flash to his wings, along with a shimmer of gray, and a dusky, shadowy twinkle, so to speak, about his head and shoulders, as you see him fluttering through the top of an orange tree or climbing, in his peculiar zigzag way, the gnarled boughs of a fig-bush. His throat and breast are the perfection of symmetry, and his eyes are clear pale gold, bright and alert. The eggs of the mocking-bird are delicate and shapely, having a body color of pale, ashy green tinged with blue and blotched with brown. The eggs of the shrike closely resemble those of the mocking-bird, so that the amateur naturalist is often deceived. The nests of the two birds are also very much alike in shape and materials, and the places in which they are usually found are exactly similar, a lonely thorny tree being preferred, if in the wildwood, and a pear-tree or a plum-tree if in an orchard.

I am quite sure that every one who has studied, or who hereafter may study, the mocking-bird in its proper haunts will agree with me that its voice is something far more marvellous than has ever been dreamed of by those who have heard it only from the cage; and especially will the lover of high dramatic art and consummate individuality of manner and vocalization be charmed with the bird's exquisite " dropping song," if once he has the good fortune to witness its delivery and hear its rhythmic gushes of rapture.

A RED-HEADED FAMILY.

"CE'TINGLY I ken, ce'tingly, seh," said my Cracker host, taking down his long flint-lock rifle from over the cabin door and slipping his frowzy head through the suspension-strap of his powder-horn and bullet-pouch. "Ce'tingly, seh, I ken cyarry ye ter wha' them air birds hed their nestis las' yer."

I had passed the night in the cabin, and now as I recall the experience to mind, there comes the grateful fragrance of pine wood to emphasize the memory. Corn "pones" and broiled chicken, fried bacon and sweet potatoes, strong coffee and scrambled eggs—a breakfast, indeed, to half persuade one that a Cracker is a *bon vivant*—had just been eaten. I was standing outside the cabin on the rude door-step. Far off through the thin pine woods to the eastward, where the sun was beginning to flash, a herd of "scrub" cattle were formed into a wide skirmish line of browsers, led by an old cow, whose melancholy bell clanged in time to her desultory movements. Near by, to the westward, lay one of those great gloomy swamps, so common in Southeastern Georgia, so repellant and yet so fascinating, so full of interest to the naturalist, and yet so little explored. The perfume of yellow jasmine was in the air, along with those indescribable woodsy odors which almost evade the sense of smell, and yet so pleasingly impress it. A rivulet, slow, narrow, and deep, passed near the front of the cabin, with a faint, dreamy murmur and crept darkling into the swamp between dense brakes of cane, and bay-bushes.

" Ye-as, seh, I ken mek er bee-line to that air ole pine snag. Hit taint more'n er half er mile out yender," continued my host and volunteer guide, as we climbed the little worm-fence that inclosed the house ; " but I allus called 'em air birds woodcocks ; didn't know 'at they hed any other name ; allus thut 'at a Peckwood wer' a leetle, tinty, stripedy feller ; never hyeard er them air big ole woodcocks a bein' called Peckwoods."

He led and I followed into the damp, môss-scented shadows of the swamp, under cypress and live-oak and through slender fringes of cane. We floundered across the coffee-colored stream, the water cooling my india-rubber wading-boots above the knees, climbed over great walls of fallen tree-boles, crept under low-hanging festoons of wild vines, and at length found ourselves wading rather more than ankle-deep in one of those shallow cypress lakes of which the larger part of the Okefenokee region is formed. I thought it a very long half-mile before we reached a small tussock whereon grew, in the midst of a dense underbrush thicket, some enormous pine trees.

" Ther'," said the guide, " thet air snag air the one. Sorter on ter tother side ye'll see the hole, 'bout twenty foot up. Kem yer, I'll show hit ter ye."

The " snag " was a stump some fifty feet tall, barkless, smooth, almost as white as chalk, the decaying remnant of what had once been the grandest pine on the tussock.

" Hello, yer' ! Hit's ben to work some more
sence I wer' yer' las' time. Hit air done dug
another hole ! "

As he spoke he pointed indicatively, with
his long, knotty forefinger. I looked and saw
two large round cavities, not unlike immense
auger-holes, running darkly into the polished
surface of the stump, one about six feet below
the other; the lower twenty-five feet above the
ground. Surely it was no very striking picture, this bare, weather-whitened column, with
its splintered top and its two orifices, and yet
I do not think it was a weakness for me to
feel a thrill of delight as I gazed at it. How
long and how diligently I had sought the home
of *Campephilus principalis*, the great king of
the red-headed family, and at last I stood before its door !

At my request, the kind Cracker now left
me alone to prosecute my observations.

" Be in ter dinner ? " he inquired as he
turned to go.

" No ; supper," I responded.

" Well, tek cyare ev yerself," and off he
went into the thickest part of the cypress.

I waited awhile for the solitude to regain its
equilibrium after the slashing tread of my
friend had passed out of hearing ; then I stole
softly to the stump, and tapped on it with the
handle of my knife. This I repeated several
times. Campephilus was not at home, for if he
had been I should have seen a long, strong,
ivory-white beak thrust out of the hole up
there, followed by a great red-crested head
turned sidewise so as to let fall upon me the
glint of an iris unequalled by that of any other
bird in the world. He had gone out early. I

should have to wait and watch; but first I satisfied myself by a simple method that my
watching would probably not be in vain. A
little examination of the ground at the base of
the stump showed me a quantity of fresh wood-
fragments, not unlike very coarse saw-dust
scattered over the surface. This assured me
that one of the excavations above was a new
one, and that a nest was either building or had
been finished but a short while. So I hastily
hid myself on a log in a clump of bushes, distant from the stump about fifty feet, whence I
could plainly see the holes.

One who has never been out alone in a
Southern swamp can have no fair understand-
ing of its loneliness, solemnity and funereal
sadness of effect. Even in the first gush of
Spring—it was now about the sixth of April—
I felt the weight of something like eternity in
the air—not the eternity of the future but the
eternity of the past. Everything around me
appeared old, sleepy, and musty, despite the
fresh buds, tassels, and flower-spikes. What
can express dreariness so effectually as the
long moss of those damp woods? I imagined
that the few little birds I saw flitting here and
there in the tree tops were not so noisy and
joyous as they would be when, a month later,
their northward migration should bring them
into our greening northern woods. As the
sun mounted, however, a cheerful twitter ran
with the gentle breeze through the bay thickets
and magnolia clumps, and I recognized a
number of familiar voices; then suddenly the
gavel of Campephilus sounded sharp and
strong a quarter-mile away. A few measured
raps, followed by a rattling drum-call, a space

of silence rimmed with receding echoes, and
then a trumpet-note, high, full, vigorous, almost startling, cut the air with a sort of broad-
sword sweep. Again the long-roll answered,
from a point nearer me, by two or three ham-
mer-like raps on the resonant branch of some
dead cypress-tree. The king and queen were
coming to their palace. I waited patiently,
knowing that it was far beyond my power to
hurry their movements. It was not long before one of the birds, with a rapid cackling
that made the wood rattle, came over my head,
and went straight to the stump, where it lit,
just below the lower hole, clinging gracefully
to the trunk. It was a superb specimen—the
female, and I suspected that she had come to
leave an egg. I could have killed her easily
with the little sixteen-gauge breech-loader at
my side, but I would not have done the act
for all the stuffed birds in the country. I had
come as a visitor to this palace, with the hope
of making the acquaintance I had so long de-
sired, and not as an assassin. She was quite
unaware of me, and so behaved naturally, her
large gold-amber eyes glaring with that wild
sincerity of expression seen in the eyes of but
few savage things.

After a little while the male came bounding
through the air, with that vigorous galloping
flight common to all our woodpeckers, and lit
on a fragmentary projection at the top of the
stump. He showed larger than his mate, and
his aspect was more fierce, almost savage.
The green-black feathers near his shoulders,
the snow-white lines down his neck, and the
tall red crest on his head, all shone with great
brilliancy, whilst his ivory beak gleamed like a

dagger. He soon settled for me a question
which had long been in my mind. With two
or three light preliminary taps on a hard heart-
pine splinter, he proceeded to beat the regular
woodpecker drum-call—that long rolling rattle
made familiar to us all by the common red-
head (*Melanerpes erythrocephalus*) and our
other smaller woodpeckers. This peculiar
call is not, in my opinion, the result of elasticity
or springiness in the wood upon which it is
performed, but is effected by a rapid, spas-
modic motion of the bird's head, imparted by a
voluntary muscular action. I have seen the
common Red-head make a soundless call on a
fence-stake where the decaying wood was
scarcely hard enough to prevent the full en-
trance of his beak. His head went through
the same rapid vibration, but no sound accom-
panied the performance. ' Still, it is resonance
in the wood that the bird desires, and it keeps
trying until a good sounding-board is found.

It was very satisfying to me when the superb
King of the Woodpeckers—*pic noir à bec blanc*,
as the great French naturalist named it—went
over the call, time after time, with grand effect,
letting go, between trials, one or two of his
triumphant trumpet-notes. Hitherto I had
not seen the Campephilus do this, though I
had often heard what I supposed to be the call.
As I crouched in my hiding-place and furtively
watched the proceedings, I remember comparing the birds and their dwelling to some half-
savage lord and lady and their isolated castle
of medieval days. A twelfth-century bandit
nobleman might have gloried in trigging himself in such apparel as my ivory-billed woodpecker wore. What a perfect athlete he ap-

peared to be, as he braced himself for an effort which was to generate a force sufficient to hurl his heavy head and beak back and forth at a speed of about twenty-eight strokes to the second !

All of our woodpeckers, pure and simple— that is, all of the species in which the woodpecker character has been preserved almost unmodified—have exceedingly muscular heads and strikingly constricted necks; their beaks are nearly straight, wedge-shaped, fluted or ribbed on the upper mandible, and their nostrils are protected by hairy or feathery tufts. Their legs are strangely short in appearance, but are exactly adapted to their need, and their tail-feathers are tipped with stiff points. These features are all fully developed in the *Campephilus principalis*, the bill especially showing a size, strength and symmetrical beauty truly wonderful.

The stiff pointed tail-feathers of the woodpecker serve the bird a turn which I have never seen noted by any ornithologist. When the bird must strike a hard blow with its bill, it does not depend solely upon its neck and head ; but, bracing the points of its tail-feathers against the tree, and rising to the full length of its short, powerful legs, and drawing back its body, head, and neck to the farthest extent, it dashes its bill home with all the force of its entire bodily weight and muscle. I have seen the ivory-bill, striking thus, burst off from almost flinty-hard dead trees fragments of wood half as large as my hand ; and once in the Cherokee hills of Georgia I watched a pileated woodpecker (*Hylotomus pileatus*) dig a hole to the very heart of an exceedingly

tough, green, mountain hickory tree, in order
to reach a nest of winged ants. The point of
ingress of the insects was a small hole in a
punk knot; but the bird, by hopping down the
tree tail-foremost and listening, located the
nest about five feet below, and there it proceeded to bore through the gnarled, cross-
grained wood to the hollow.

Of all our wild American birds, I have
studied no other one which combines all of the
elements of wildness so perfectly in its character as does the ivory-billed woodpecker. It
has no trace whatever in its nature of what
may be called a tamable tendency. Savage
liberty is a prerequisite of its existence, and its
home is the depths of the woods, remotest
from the activities of civilized man. It is a
rare bird, even in the most favorable regions,
and it is almost impossible to get specimens of
its eggs. Indeed, I doubt if there are a dozen
cabinets in all the world containing these eggs;
but they are almost exactly similar in size,
color and shape to those of *Hylotomus pileatus*,
the only difference being that the latter are,
upon close examination, found to be a little
shorter, and, as I have imagined, a shade less
semi-transparent porcelain-white, if I may so
express it.

The visit of my birds to their home in the
stump lasted nearly two hours. The female
went into and out of the hole several times
before she finally settled herself, as I suppose,
on her nest. When she came forth at the end
of thirty or forty minutes, she appeared exceedingly happy, cackling in a low, harsh,
but rather wheedling voice, and evidently
anxious to attract the attention of the male,

who in turn treated her with lofty contempt.
To him the question of a new egg was not
worth considering. But when she at last
turned away from him, and mounting into the
air, galloped off into the solemn gloom of the
cypress wood, he followed her, trumpeting at
the top of his voice.

Day after day I returned to my hiding-place
to renew my observation, and, excepting a
visitation of mosquitoes now and then, nothing occurred to mar my enjoyment. As the
weather grew warmer the flowers and leaves
came on apace, and the swamp became a vast
wilderness of perfume and contrasting colors.
Bird songs from migrating warblers, vireos,
finches and other happy sojourners for a day
(or mayhap they were all nesting there, I cannot say, for I had larger fish to fry), shook the
wide silence into sudden resonance. Along
the sluggish little stream between the canebrakes, the hermit-thrush and the cat-bird were
met by the green heron and the belted kingfisher. The snake-bird, too, that veritable
water-dragon of the South, was there, wriggling and squirming in the amber-brown pools
amongst the lily-pads and lettuce.

At last, one morning, my woodpeckers discovered me in my hiding-place; and that was
the end of all intimacy between us. Thence-
forth my observations were few and at a long
distance. No amount of cunning could serve
me any turn. Go as early as I might, and hide
as securely as I could, those great yellow eyes
quickly espied me, and then there would be a
rapid and long flight away into the thickest
and most difficult part of the swamp.

I confess that it was with no little debate

that I reached the determination that it was my duty to rob that nest in the interest of knowledge. It was the first opportunity I ever had had to examine an occupied nest of the *Campephilus principalis*, and I felt that it was scarcely probable that I should ever again be favored with such a chance. With the aid of my Cracker host, I erected a rude ladder and climbed up into the hole. It was almost exactly circular, and nearly five inches in diameter. With a little axe I began breaking and hacking away the crust of hard outer wood. The cavity descended with a slightly spiral course, widening a little as it proceeded. I had followed it nearly five feet when I found a place where it was contracted again, and immediately below was a sudden expansion, at the bottom of which was the nest. Five beautiful pure white eggs of the finest old-china appearance, delicate, almost transparent, exceedingly fragile, and, to the eyes of a collector, vastly valuable, lay in a shallow bowl of fine chips. But in breaking away the last piece of wood-crust, I jerked it a little too hard, and those much coveted prizes rolled out and fell to the ground. Of course they were "hopelessly crushed," and my feelings with them. I would willingly have fallen in their stead, if the risk could have saved the eggs. I descended ruefully enough, hearing as I did so the loud cry of Campephilus battling around in the jungle. Once or twice more I went back to the spot in early morning, but my birds did not appear. I made minute examination of the rifled nest, and also tore out the other excavation, so as to compare the two. They were very much alike, especially in the

jug-shape of their lower ends. From a careful study of all the holes (apparently made by
Campephilus) that I have been able to find
and reach in either standing or fallen trees, I
am led to believe that this jug-shape is peculiar to the ivory-bill's architecture, as I have
never found it in the excavations of other
species, save where the form was evidently the
result of accident. The depth of the hole
varies from three to seven feet, as a rule, but I
found one that was nearly nine feet deep and
another that was less than two. Our smaller
woodpeckers, including *Hylotomus pileatus*,
usually make their excavations in the shape of
a gradually widening pocket, of which the en-
trance is the narrowest part.

It is curious to note that—beginning with the
ivory-bill and coming down the line of species
in the scale of size—we find the red mark on
the head rapidly falling away from a grand
scarlet crest some inches in height to a mere
touch of carmine, or dragon's blood, on crown,
nape, cheek, or chin. The lofty and brilliant
head-plume of the ivory-bill, his powerful beak.
his semi-circular claws and his perfectly spiked
tail, as well as his superiority of size and
strength, indicate that he is what he is, the
original type of the woodpecker, and the one
pure species left to us in America. He is the
only woodpecker which eats insects and larvæ
(dug out of rotten wood) exclusively. Neither
the sweetest fruits nor the oiliest grains can
tempt him to depart one line from his hereditary habit. He accepts no gifts from man, and
asks no favors. But the pileated woodpecker,
just one remove lower in the scale of size,
strength, and beauty, shows a little tendency

towards a grain and fruit diet, and it also often
descends to old logs and fallen boughs for its
food—a thing never thought of by the ivory-
bill. As for the rest of the red-headed family,
they are degenerate species, though lively,
clever, and exceedingly interesting. What a
sad dwarf the little downy woodpecker is when
compared with the ivory-bill! and yet to my
mind it is clear that *Picus pubescens* is the degenerate off-shoot from the grand *campephilus*
trunk.

Our red-headed woodpecker (*M. erythrocephalus*) is a genuine American in every sense,
a plausible, querulous, aggressive, enterpris-
ing, crafty fellow, who tries every mode of getting a livelihood, and always with success. He
is a woodpecker, a nut-eater, a cider-taster, a
judge of good fruits, a connoisseur of corn,
wheat, and melons, and an expert fly-catcher as
well. As if to correspond with his versatility
of habit, his plumage is divided into four regular masses of color. His head and neck are
crimson, his back, down to secondaries, a
brilliant black, tinged with green or blue in
the gloss ; then comes a broad girdle of pure
white, followed by a mass of black at the tail
and wing-tips. He readily adapts himself to
the exigencies of civilized life. I prophecy
that, within less than a hundred years to come,
he will be making his nest on the ground, in
hedges or in the crotches of orchard trees.
Already he has begun to push his way out into
our smaller Western prairies, where there is no
dead timber for him to make his nest-holes in.
I found a compromise-nest between two fence-
rails in Illinois, which was probably a fair index
of the future habit of the red-head. It was

formed by pecking away the inner sides of two
vertical parallel rails, just above a horizontal
one, upon which, in a cup of pulverized wood,
the eggs were laid. This was in the prairie
country between two vast fields of Indian corn.

The power of sight exhibited by the red-
headed woodpecker is quite amazing. I have
seen the bird, in the early twilight of a summer
evening, start from the highest spire of a very
tall tree, and fly a hundred yards straight to an
insect near the ground. He catches flies on
the wing with as deft a turn as does the great-
crested fly-catcher. It is not my purpose to
offer any ornithological theories, in this paper ; but I cannot help remarking that the farther a species of woodpecker departs from the
feeding-habit of the ivory-bill, the more broken
up are its color-masses, and the more diffused
or degenerate becomes the typical red tuft on
the head. The golden-winged woodpecker
(*Colaptes auratus*), for instance, feeds much on
the ground, eating earth-worms, seeds, beetles,
etc. ; and we find him taking on the colors of
the ground-birds with a large loss of the characteristic woodpecker arrangement of plumage
and color-masses. He looks much more like
a meadow-lark than like an ivory-bill ! The
red appears in a delicate crescent, barely noticeable on the back of the head, and its bill
is slender, curved, and quite unfit for hard
pecking. On the other hand, the downy
woodpecker, and the hairy woodpecker, having
kept well in the line of the typical feeding
habit, though seeking their food in places be-
neath the notice of their great progenitor,
have preserved in a marked degree an outline
of the ivory-bill's color-masses, degenerate

though they are. The dwarfish, insignificant looking *Picus pubescens* pecking away at the stem of a dead iron-weed to get the minute larvæ that may be imbedded in the pith, when compared with *Campephilus principalis* drumming on the bole of a giant cypress-tree, is like a Digger Indian when catalogued in a column with men like Goethe and Gladstone, Napoleon and Lincoln.

I have been informed that the ivory-bill is occasionally found in the Ohio valley ; but I have never been able to discover it north of the Cumberland range of mountains. It is a swamp bird, or rather it is the bird of the high timber that grows in low wet soil. Its principal food is a large flat-headed timber-worm known in the South as *borer* or *saw-worm*, which it discovers by ear and reaches by diligent and tremendously effective pecking. A Cracker deer-stalker, whom I met at Blackshear, Georgia, gave an amusing account of an experience he had had in the swamps. He said :

" I had turned in late, and got to sleep on a tussock under a big pine, an' slep' tell sun-up. Wull, es ther' I laid flat er my back an' er snorin' away, kerwhack sumpen tuck me in the face an' eyes, jes' like spankin' er baby, an' I wuk up with er gret chunk er wood ercross my nose, an' er blame ole woodcock jest er whangin' erway up in thet pine. My nose hit bled an' bled, an' I hed er good mint er shoot thet air bird, but I cudn't stan' the expense er the thing. Powder'n' lead air mighty costive. Anyhow I don't s'pose 'at the ole woodcock knowed at hit'd drapped thet air fraygment onto me. Ef hit'd er 'peared

like's ef hit wer' 'joyin' the joke any, I wud er
shot hit all ter pieces ef I'd er hed ter lived
on turpentime all winter ! "

Of the American woodpecker there are more
than thirty varieties, I believe, nearly every
one of which bears some trace of the grand
scarlet crown of the great ivory-billed king of
them all. The question arises—and I shall
not attempt to answer it—whether the ivory-
bill is an example of the highest development,
from the downy woodpecker, say, or whether
all these inferior species and varieties are the
result of degeneracy? Neither Darwin nor
Wallace has given us the key that certainly
unlocks this very interesting mystery.

The sap-drinking woodpeckers (*Sphyropicus*),
of which there are three or four varieties in
this country, appear to form the link between
the fruit-eating and the non-fruit-eating species
of the red-headed family. From sipping the
sap of the sugar-maple to testing the flavor of
a cherry, a service-berry, or a haw-apple, is a
short and delightfully natural step. How logi-
cal, too, for a bird, when once it has acquired
the fruit-eating habit, to quit delving in the
hard green wood for a nectar so much inferior
to that which may be had ready bottled in the
skins of apples, grapes, and berries ! In accordance with this rule, *M. erythrocephalus*
and *Centurus carolinus*, though great tipplers,
are too lazy or too wise to bore the maples,
preferring to sit on the edge of a sugar-trough,
furtively drinking therefrom leisurely draughts
of the saccharine blood of the ready-tapped
trees. I have seen them with their bills
stained purple to the nostrils with the rich
juice of the blackberry, and they quarrel

from morning till night over the ripest June-
apples and reddest cherries, their noise making a Bedlam of the fairest country orchard.

The woodpecker family is scattered widely
in our country. In the West Canadian woods
one meets, besides a number of the commoner
species, Lewis' woodpecker, a large, beautiful,
and rare bird. The California species include
the Nuttall, the Harris, the Cape St. Lucas, the
white-headed, and several other varieties, all
showing more or less kinship to the ivory-bill.
Lewis's woodpecker shows almost entirely
black, its plumage giving forth a strong greenish or bluish lustre. The red on its head is
softened down to a fine rose-carmine. It is
a wild, wary bird, flying high, combining in
its habits the traits of both *Hylotomus pileatus*
and *Campephilus principalis.*

In concluding this paper a general description of the male ivory-bill may prove accept-
able to those who may never be able to see
even a stuffed specimen of a bird which, taken
in every way, is, perhaps, the most interesting
and beautiful in America. In size 21 inches
long, and 33 in alar extent; bill, ivory white,
beautifully fluted above, and two and a-half
inches long; head-tuft, or crest, long and
fine, of pure scarlet faced with black. Its
body-color is glossy blue-black, but down its
slender neck on each side, running from the
crest to the back, a pure white stripe contrasts
vividly with the scarlet and ebony. A mass
of white runs across the back when the wings
are closed, as in *M. erythrocephalus*, leaving the
wing-tips and tail black. Its feet are ash-
blue, its eyes amber-yellow. The female is
like the male, save that she has a black crest

instead of the scarlet. I can think of nothing in Nature more striking than the flash of color this bird gives to the dreary swamp-landscape, as it careers from tree to tree, or sits upon some high skeleton cypress-branch and plies its resounding blows. The species will probably be extinct within a few years.*

* Since writing the foregoing, I have made several excursions in search of the ivory-bill. Early in January, 1885, I killed a fine male specimen in a swamp near Bay St. Louis, Mississippi; but was prevented, by an accident, from preserving it or making a sketch of it.

TANGLE–LEAF PAPERS.

I.

In the season of nest-building, which is also the season of song-singing, the by-ways of American rural districts offer many attractions to the student of nature, and especially to the student who hopes to turn his discoveries to account in any field of art. Of mere descriptive matter, so far as it may go in literature, and of mere conventionalization, so far as decorative drawing and painting are concerned, the most that was ever possible has, probably, already been done ; but the higher forms of art, which we have agreed to call creative, must get the germs of all new combinations from the suggestions of nature. I often have thought that even criticism in our country would have more virility in it if the critics had more time and more inclination to study nature outside of cities and greenhouses. How can Wordsworth be studied with true critical insight by one who but vaguely remembers the outlines of the woods and fields, the shady lanes, and the fine aerial effects of hilly landscape ? When one with open eyes and ears goes out into the unshorn ways of nature in the creative season—spring—the fine fervor at work in birds, and trees, and plants, in the air, the earth, and the water, is so manifest that one cannot doubt that some subtle element of originality is easily obtainable therefrom by infection. Of course one must be susceptible to

the most delicate shades of influence in order
to get the values of nature. Even the photo-
graph is to be caught on no plate save the most
sensitive.

The other day, when I told a friend that I
had discovered that the mocking-bird never
tries to imitate the cooing of a dove, he said,
"Why, every one knew that long ago."—
"Show me the record," I demanded ; but he
could not. "Well, what good can come of
your discovery, even if you are entitled to the
credit?" he rather triumphantly asked. I
answered that the fact was suggestive ; that it
had an artistic value. A mournful, desponding
voice is never attractive to a vigorous, healthy
nature. Cheerfulness and enthusiasm are
what win followers for birds as well as men.
The mocking-bird is a genius who catches from
nature all its available notes, and combines
them so as to express the last possibility of
bird-song, rejecting the moaning of the dove
and the thumping notes of the yellow-billed
cuckoo, just as the true poet rejects thoughts
and words unworthy of his lay.

It is true that, as the times go, the artist is
called upon to please a vitiated taste. The
poet and the novelist must meet the demands
of the schools and coteries. The precious
hints and suggestions caught from the provincial lanes and wood-paths are not considered
favorable by the metropolitan, as a rule ; but
out of these must grow, as the plant from the
seed, the living, lasting values of all art. City
study is book study, through which the truths
and beauties of nature are seen at a distance,
as if through a very delusive atmosphere. To
test this take your books into the woods of

spring, beside a brook, and see how many of them will bear reading in the light and presence of nature. How tasteless become the polished bits of conventional art when we attempt to enjoy them in the open air, where the violets grow, and the wild vine hangs its festoons!

There is another test of the force and vitality of nature's suggestions known to every observant artist. For instance, a sketch of some out-door scene, made on the spot, will appear to have scarcely any value so long as it can be readily compared with the original; but no sooner is the portfolio opened in the studio than the sketch discloses, in a marked degree, many of the subtlest beauties or peculiarities of the living scene. How different in the case of a sketch made from the flat! How diluted the power of nature becomes!

I was once enjoying a luncheon with a gay sylvan party, when the earth served as table and a sward of blue-grass as table-cloth. A lady who gloried in her collection of rare hand-painted china was serving tea to us in cups worth more than their weight in gold; and yet when one of these chanced to be set down in the midst of a tuft of wild violets it was so dulled by contrast with the living blooms that it really appeared coarse and crude. To study nature is the surest way to a knowledge of what art ought to be. Nature is the standard. I have little respect for the judgment of the critic who measures one man's work by that of another. The main question, when any art-work is to be critically considered should be, Has it the symmetry, force, and vital beauty of nature?

It is easy to write about nature ; but to write in the spirit of nature, to keep within the limit of her rules, is not so easy. So to copy all the salient features of a landscape is within the power of any painter, but how few can get their brushes to spill upon the canvas even a modicum of what we all may see in the sky, and sea, and shore ! Greening hedgerows, and blooming orchards, the songs of the cat-bird and brown thrush, always have something new in them. We never see or hear them twice from the same point of observation. The brook's voice has an infinite variety of tones. The sunlight and the cloud shadows are continually changing. And so if one can hoard up the impressions made by the thousand passing moods of spring, they will prove richly suggestive when reviewed in the quiet of the study. The fine mass of such impressions will be found a fresh and fragrant matrix, enclosing the perfect crystals of original thought. If it is true that one grows like what one contemplates nothing but good can come of lonely rambles with nature, and especially in the season of quickening germs and tender impulses.

Those who assert that there is nothing especially picturesque or strikingly interesting in our rural scenery seem to me deficient either in judgment or in the power of observing closely. The fact is, it is hard for the professional artist or literary man to cut loose from an hereditary old-country taint. The far-away, the dim, the old in literature and art are shrouded in the blue enchantment that hovers so tantalizingly on all heights. Standing on one mountain-top we look to another longingly ; reclining on one bank of a river we dream of

the joy awaiting us on the other. It is, in
other words, apparently almost impossible for
Americans to fully recognize and appreciate
the richness of "local color" everywhere offered at home.   If we knew our country as well
as the English know theirs we should have a
stronger vital energy in our literature and art.
Of course we lack that long perspective and rich
historical atmosphere belonging to old countries, but as a nation we are just at that age
when our genius should find its note.   Our
highways are reasonably good, our lanes and
by-ways are inviting, our people are hospitable
and communicative.   There is no good reason
why some tourists, of a more interesting sort
than tax-gatherers and lightning-rod peddlers,
should not explore the pastoral districts where
the richest materials for poetry, romance, and
art may be had for the taking.

Rummaging the remote nooks of literature—
the pages of Chaucer and Spenser, and Izaak
Walton and Roger Ascham, or François Villon
and Marot and Ronsard, is very pleasing and
profitable ; but the living, budding, redolent,
and resonant by-ways of our own neighborhoods
offer a richer reward.   There are moments
when there are a fragrance and savor, so to
speak, in the song of a plough-boy heard across
the fresh-turned fields.   One pauses by the
fence or hedge-row to enjoy what no book or
picture can quite give.   A breath of perfume
from the blooming top of a wild crab-apple tree,
along with the hum of the bees at work there,
is a poem much older than any ballade or triolet, and fresher and sweeter than any song of
troubadour or any idyl of Greek lyrist.   What
matters it whether one walks, or rides a tri-

cycle, or spins noiselessly along on a bicycle, so that one keeps one's eyes and ears open? If the body is to be refreshed and strengthened by exercise, why not also take pains to recreate the mind by filling the memory with pungent and healthful data? A cool draught from a country way-side spring, where the calamus grows, and the little platoons of sky-blue butter-flies arrange themselves on the damp spots, might well inspire an ode as good as any Ana-creon ever drew from the purple grape-juice. The first dragon-fly of the season is always a happy discovery for me.

I know where Longfellow got the sugges-tions for his *Flower de Luce*, the fresher stanzas, at least; for the dew of morning, brushed from brook-side flags and meadow weeds, is in them. The poem is bookish, too, showing the scholar a little too plainly, perhaps; but it serves to urge a current of out-door air over one as one reads, and the sound of the mill-flume is in the measure. It is always a charming junction where ripe scholarship and an accurate and loving knowledge of nature flow together. From that point onward how the imagination is enriched!

The poems of Theocritus and the song of the cardinal-bird are blended together, and something new comes of the mixture. I like to follow through a racy poem or essay some elu-sive, fascinating trace of the author's recipe. It is never quite hidden.

The impetus given to out-door rambling by the advent of cycling must, it seems to me, bring some fresh elements into American thought. It will, unless we allow the love of mere whirling to shut out everything else. I

have found a tricycle the most helpful and en-
joyable thing in exploring the by-ways and
high-ways of my neighborhood.   It has helped
me to see things that I might not have discovered had I been on foot, and it has awakened
sensations never before experienced by me.
The mere joy in self-propulsion seems to
sharpen one's vision, and strengthen one's re-
ceptive faculties.   I like to stop and sit in the
saddle, and peep between the rails of a fence,
letting my eyes follow the fresh green rows of
young Indian corn that reach far across the
level field of dark loam.   From the same po-
sition I can make such notes and sketches as
will be of use to me in the future.   Charming
physical exercise and pleasing study combined
make up about the most desirable of all com-
pounds.   When I am tired of pedalling I can
stop in the shade of a way-side tree and draw
forth a book to read, or I can watch the effect
of cloud-shadows and wind-flaws on the rank
green wheat.   Meadow-larks and blue-birds
preen themselves on the fence-stakes, field-
sparrows sing in the young oats, yonder orchard rings with the medley of the cat-bird.
Here is a good place to test the qualities of a
book as an out-door companion.   One can find
out how its pages will accord with certain
phases of nature, so to speak.   Ten to one what
had seemed quite perfect, read in the atmosphere of the library, will fall off to a mere skeleton in the open air.   I have found that,
strange as it may seem, the poems of Burns
lose something by out-door reading, whilst cer
tain passages of Tennyson, Browning, and
Emerson reach out and gather an increment of
freshness from pastoral surroundings.   The

humorists, as a rule, require to be read within the limitations of four walls. Nature is always in earnest.

A novel that will bear the sunlight and the winds and the bird-songs may be put down as a thoroughly good one. Short, crisp stories, not too tragic, having strong local color and bright conversations, stand this test very well. Our magazines often fall into the error of printing, during the out-door season, light society stories of city life; these fade into colorless and tasteless films when read on the beach, or in the open country. I sometimes read French novels out-of-doors, merely for the antiseptic effect that the sun and air have on the offensive passages; but at best I often find myself glad that American birds and flowers do not understand French.

We Americans are too fast with whatever we undertake. Our horses must trot "below fifteen," our yachts must go like a hurricane; and when we ride bicycles or tricycles we must run a hundred miles in the shortest possible space of time. Now, a tourist who hopes to see anything or hear anything worth remembering must go slowly over his ground, with many stops and with all sorts of detours. One never can foreknow what odd and interesting things may be discovered tucked away in unfrequented nooks. I have experienced many pleasing surprises in the way of valuable information drawn from most unpromising sources. Such rich dialect phrases, too, and such rare, quaint traits of character, disclose themselves! How marvellously weatherwise some of the country folk are, and what keen observers of nature! On the other hand, they

have such queer "notions" about signs and
omens.   For instance, the well-known guttural
croaking of the yellow-billed cuckoo is, in the
· West and South generally believed to presage
rain; hence the bird is known amongst the
rural people by the name of rain-crow.

I remember with what solemn earnestness
an old man once heaped maledictions on a
cuckoo. It was in the midst of a distressing
drought, and the bird was mournfully uttering
its notes in an orchard. "There's thet air
dad-blasted rain crow a-bellerin' down ther'
ag'in" he cried, savagely wagging his head.
"Ef I hed a gun I'd blow it inter thunder 'n'
gone. Ever'thin' a-burnin' up an' the crick a-
goin' dry an' thet air lyin' rain-crow jest
a-yowkin' an' yowkin', es ef a flood wer' a-
comin' in less an' fifteen minutes—blast its
pictur'!"

Speaking of the yellow-billed cuckoo, it is
one of the most interesting of our American
birds,—a late comer to our Northern woods,
where about the middle of May it begins a
shy, shadowy pilgrimage from tree to tree,
peering furtively among the tufts of young
leaves, as if bent on some errand of mystery.
It is a slender, graceful figure, with a dispro-
portionately long tail and a slim, slightly curved
bill, which is almost black above and yellow
below; its back is drab; its under parts a pure
silvery-white, and its tail dark, tipped with
snow-white. You may know it by its peculiar
zigzag flight, and by its cry, "*Kaow, Kaow,*"
etc., repeated slowly at first, then increasing in
rapidity to a rattling or pounding croak, and
finally ending laggingly as it began. It has all

the most interesting habits of the English cuckoo.

I am aware that naturalists have stoutly claimed that our yellow-bill never lays its eggs in other birds' nests , but I have the evidence of my own eyes to the contrary. I was plying a country lad with questions touching the birds and nests of his neighborhood, when he informed me that a robin and a rain-crow had a nest " in cahoot " * in an apple-tree just across a lane from where we stood. Of course I was anxious to see that nest at once. It was built in the usual robin fashion, stacked up in a low crotch of the tree, and contained three robin eggs and one cuckoo egg. This was a number of years ago.; but so late as the spring of 1883 I found a cuckoo's egg in the nest of a blue-jay. In the mountain region of North Georgia, where the yellow-bill nests among the haw thickets, I have seen it carrying its egg in its mouth, no doubt with the purpose of depositing it in the care of some other bird. Wherever I have gone I have heard this cuckoo charged with eating the eggs of other birds; but I believe the charge has no better foundation than the mistake of observers, who, seeing it with its own egg in its mouth, naturally suppose that it has been robbing some neighborbird's nest. My opinion is, that by the time our country shall have reached the age of the England of to-day our cuckoos will have become confirmed in all the habits of the European species. At best the bird is very indifferent to nest-building, and its natural bent is towards entirely evading the reponsibility.

* " In cahoot " is a common Western and Southern phrase for in partnership.

4

Its architectural powers are of the poorest.
No other of our arboreal birds, not even the
common dove, builds so crazy and insecure a
home.   But I am getting into rather deep ornithological mire.   It is so easy to find room
for digression when one gets out-of-doors !
Everything is suggestive.   To the vision of a
careful observer and student each object in nature has an interrogation-point beside it.   With
pencil and note-book let us catalogue these
suggestions and interrogations, and lay them
aside for future use.   When, some day, we
come to look them over we shall be surprised
how perfectly—like dried roots and plants—
they have kept their out-door fragrance and
taste.

II.

In studying the birds most usually met with
on out-door excursions I have found it very interesting to make notes of certain striking evi-
dences of a special harmonic relation between
their movements, colors, and attitudes, and
the peculiarities of their natural surroundings.
Ornithologists have over and over again
rung the changes on the ease with which the
quail, the grouse, and the hare make themselves next to invisible to the human eye, and
to the piercing vision of birds of prey as well ;
but there are many curious details connected
with this subject of a natural harmony of motion and color, regarding birds and their envi-
ronments, which I have never seen in print.
Of course, since the quail, the hare, and the
grouse have been for so long the objects of
desire of sportsmen, pot-hunters, and epicures,

as well as of careful study by naturalists, their peculiarities have all been catalogued, and every intelligent person knows that a hare, by crouching flat on a dry gray spot of earth, so blends with its surroundings as to become almost undistinguishable, and that a quail, sitting in a handful of dry brown leaves is as effectually hidden as if buried. So a grouse among the tangled twigs of a bare winter tree is a very difficult object to discover. A meadow-lark, in a sunny clover-field, melts, so to speak, into the general confusion of brown, green, and gold, so that it becomes indeed a "sightless song." The humming-bird makes its nest of lichen, and places it in a tuft of the same on some wrinkled bough, usually at or near a crotch ; and the little bird, while on the nest, is so in harmony with its surroundings that none but the keenest eye would distinguish her from one of the little ruffled knots on the bark beside her. The whippoorwill builds no nest. Its eggs are deposited·on the ground at a place where the bird's colors and those of her eggs perfectly harmonize with the general tone of their surroundings. I have known this bird to roll her eggs from spot to spot while incubating, evidently for the purpose of keeping them and herself within a proper *entourage*, this being her only means of protection from hawks, owls, and other enemies. The common dove places its shallow, ill-made nest in what appear to be the most exposed places, but the bluish ash-gray color of the bird's plumage runs so evenly into the tone of its surroundings that one might look in vain for any sign of a living thing in the midst of that apparently flat wash of drab neutral.

That hawks and owls have powerful and far-seeing eyes cannot be doubted; but they either lack a fine power of discrimination in vision, or this adaptation of the colors and markings of birds to their surroundings is very effectual, else these birds of prey exhibit a wonderful forbearance toward their natural victims during the season of incubation. I am inclined to the opinion that hawks are what might be called "far-sighted," and that their vision at very short distances is not very clear. I once saw a goshawk pursuing a downy woodpecker, when the latter darted through a tuft of foliage and flattened itself close upon the body of a thick oak bough, where it remained as motionless as the bark itself. The hawk alighted on the same bough within two feet of its intended victim, and remained sitting there for some minutes, evidently looking in vain for it, with nothing but thin air between monster and morsel. The woodpecker was stretched longitudinally on the bough, its tail and beak close to the bark, its black and white speckled feathers looking like a continuation of the wrinkles and lichen. No doubt those were moments of awful suspense for the little fellow; but its ruse succeeded, and the hawk flew away to try some other tidbit. If the woodpecker had stopped amongst the green leaves, the hawk would have discovered it instantly.

I have noticed that the cardinal-grosbeak and the blue-jay are more often killed by hawks than are the other common birds of our woods; and I attribute the fact to their brilliant plumage. The blue-jays are aware of their danger, and resort to mob-law whenever a hawk or owl is discovered. I have seen a hundred blue-jays bonded together and worrying one little screech-owl. The grosbeaks protect themselves as best they can by keeping well within thickets and thorny close-topped trees.

Along our rivers and brooks live a great many aquatic and semi-aquatic birds, whose traits and peculiar characteristics seem not to have been very closely noted by our naturalists.

I have mentioned the motions and attitudes of birds as partaking of the general tone of their surroundings. This is particularly observable in the herons, sand-pipers, plovers, bitterns, and many shore birds. The motionless, dreamy appearance of the heron as it stands in the edge of a still gray pool of water is in perfect keeping with all the features and accessories of a tarn. So the wavering, tilting motion of the little sand-pipers accords harmoniously with the rippling surface of running water. So accentuated is this light see-saw movement of one of the lesser sand-pipers, that the bird is called "teeter-snipe" by the country folk. The kill-deer plover, common in our damp meadows and fallow lands, has a way of running in the low grass and stubble that renders it very hard to follow with the eye, and, when it stops, its outlines are so shadowy and so intimately blent with the gray-brown background that one has to look sharply to discover it. The little green heron of our brooks and rivulets has a habit of sitting on old heaps of drift-wood, where he looks for all the world like an upright stick or piece of bark. When

standing in the water his colors shade off into
the greenish wash of the stream, and you rarely
see him, no matter how near him you may be,
before he springs into the air, and is away. I
once shot a fine specimen as it flew past me,
and it fell among some stones at a brook's
edge. Something attracted my eyes from the
spot where it fell, and when I turned again to
look for my bird I could not see it. I walked
round and round. I knew it had fallen quite
dead ; but what had become of it ? In fact it
lay there in plain view under my eyes ; but its
colors were so uniform with those of the
smooth, water-washed stones, amongst which
it had fallen, that I was full five minutes discovering it. Every sportsman has experienced
similar difficulty in looking for snipe and woodcock after bringing them down.

The kingfisher's colors are, no doubt, of
great advantage to him in taking his prey from
the water. If he were red, instead of being
dashed over with all the blue and purple and
silver-gray, and liquid shadows of the brook
itself, he would not catch many fish. How
hard it must be for the minnows, as they dis-
port in the dancing current, to see, through the
trembling medium, the sky-blue and silvery
markings of the bird sitting on a swaying
branch between them and the sky ! And how
easy it would be for the kingfisher to get all
the food he might desire if those little fish were
less of the color of the water in which they
swim. If quails were scarlet instead of mottled
brown, how soon the hawks would exterminate
them !

But there is another side to this subject of
which the poet and artist must take careful

note. Nature's tone is rarely loud, rarely over-accentuated. The blue-jay in the orchard, the cat-bird in the hedgerow, the kingfisher by the brook, each is a key to a harmony. Nature, on the whole, suggests under-statement and a reserve of color. Her contrasts are not of the Rembrandt type; her expressions do not abound in adjectives. Gay, flaunting flowers and gorgeous birds are rare save in greenhouses and cages. The suppressed power felt in the solemn stillness of great woods is suggestive of that force which some men of few words bear about with them.

I saw a simple picture of Nature's painting once, which has returned to my memory again and again, and if it could be put on a canvas or fastened in a poem it would forever remain a masterpiece of art. And yet it was nothing but a green heron standing in the swift shallow current of a brook with the diamond-bright wavelets breaking around its slender legs and a tuft of water-grass trembling beside it. I was lying idly enough, at full length on the brook's bank, so that beyond the bird, as I gazed, opened a fairy-like landscape, over which a gentle breeze was blowing with an effect wholly indescribable, shaking tall flags and tossing the dragon-flies about in the sunshine. The whole effect was cooling and tranquillizing, with a subtle hint in it of a land somewhere just out of reach where one might dream the lotos-dream forever.

Now, a good artist might have easily painted the little scene so far as painting usually goes; but it would have required such genius as is yet to be born to imprison in the sketch the hint of what seemed to lie just beyond the

dreamy horizon. None but the most master-
ful genius would have been able to keep up to
the sweet, quiet key of the coloring, and yet be
satisfied with the tender, wavering outlines and
the soft, transparent shadows. The liquid
tones of sound and color in the brook came so
harmoniously to my senses, along with the
motion of swaying flags and bubble-beaded
waves, that the graceful bird, seen through
half closed eyes, appeared to be a half-fanciful
embodiment of the spirit of calm delight, knee-
deep in some tide of enchantment or romance.
(Looking back over this last sentence I recognize its weakness, but must wilfully let it go,
for it comes very near expressing one phase of
the view.) Nature is rarely either *flamboyant*
or *grisâtre*, but keeps well the golden mean.

But, to return to the motions of birds, how
perfectly in keeping with the broad expanse of
sky and the movements of the clouds is the
sailing of the great-winged hawks and vultures!
I have watched the swallow-tailed hawks of
the South sailing so high that they appeared to
be sliding against the sky. No labored movements there; those wings were far above the
difficulties that beset our earth, and were
spread on heavenly tides. Even the obscene
turkey-buzzard, when it has reached a great
altitude, and is moving so smoothly and
dreamily between us and the empyrean, becomes an object of respect; we forget its vul-
garity, as we do that of men who have mounted
on the wings of genius, bearing their depravi-
ties into the rare atmosphere of exalted art.
The albatross, that prince of the sea-winds,
seems a part of the fleece-clouds and the sky.
The flamingoes, the pelicans, the gulls—all

the wild sea-fowl and shore-birds have something of the ocean-swell and the surf-ripple in
their flight. I believe it is Dr. Holmes who
speaks of the

"Oriole floating like a flake of fire,"

but, true as the comparison is, the oriole, with
its sunshine and shadows, harmonizes perfectly
with the fresh greens and yellows of the young
spring leaves and tassels. How many of our
fly-catchers, finches, and warblers have a dash
of sap green and pale leaf-yellow, as if Nature
had purposely meant them for a part of her
general spring scheme of color! Even the
bull-frog has the same marking as the tuft of
water-grass in which he sits ready for his head-
long plunge into the pool. Need I remind the
experienced sportsman of the fact that a wood-
duck among the broad leaves and snowy
blooms of the water-lily is a thing almost impossible to see although in plain view? The
beautiful bird's white and gray and purplish
markings blend easily with the water-gleams,
and leaf-shimmer, and pure white flower-clus-
ters.

The herons and kingfishers have for ages
set an example that anglers have not had the
wit to follow. White and pale blue are the
water high-lights as seen from *under the surface
of the water.* A white coat, with misty, dark-
gray wading-boots, would be nearly the snowy-
heron's fishing outfit for still, murky water.
Why? Because the legs must be in the water
and the coat above water. So the great blue-
heron has dark gray-brown legs, and all its
under parts are overlaid with fine narrow
feathers of silvery white. But the kingfisher,

whose prey is taken from clear, moving water—
are peculiarly marked underneath. On his
breast, next to his white necklace, is a band of
pale blue, touched here and there with light-
brown, and below this to his tail he is white.
Now, a fish looking up through the water has
the kingfisher between him and the sky.
Those sky-blue and silver-white feathers correspond exactly with the water-light and sky-
light as they are broken up and blended together by the tiny chopping waves. When the
kingfisher makes a harpoon of itself, and, beak
downward, darts from its perch above the
water to fall upon a fish, it presents two parallel curved lines, one of which is mainly
bright blue, the other mostly pure white ; these
seen through moving water blend into a soft
mist-gray, perfectly in tone with the prevailing
tint of most brook-water.

In connection with observations on the motions of birds it is well to recall the fact that
nearly all the night-birds fly on wings that
make no sound. An owl slips through the air
with the utter silence of a shadow. This accords with the stillness of the night. It also
serves the bird a good turn, for the least noise
would startle his prey at a time when all nature
is hushed and breathless. I have observed,
as has every nature-student, I suppose, that
nearly, if not quite all, the night insects are
comparatively noiseless in their flight. The
giant moth does not hum like a bumble-bee or
a humming-bird. The mosquito is the noisiest
with his wings of all the night-flyers. But I
must not get over the line from birds to insects, while on this subject of harmony, for a
study of butterflies alone would fill more space

than I have for this paper. In tropical and semi-tropical countries a curious resemblance in color and shape exists between the butter-flies and the flowers they haunt, a resemblance quite noticeable as far north as the fortieth degree of latitude.

III.

How would " Tricycles and Triolets " do for an alliterative heading to a light chapter on out-door poetry? Ever since I began to taste Virgil in my school-days I have had a special liking for verse smacking of the woods and fields, the birds, the sunshine, and the brooks. A certain passage in the Æneid comes into my mind now, a strong sketch of a grove of trees, with the light playing through the swaying foliage with that strangely brilliant effect so often observed on bright days in spring and summer :—

—" Tum silvis scena coruscis
Desuper, horrentique atrum nemus imminet umbrâ."

I do not think that William Morris has quite done justice to this beautiful Virgilian bit of landscape in his rhymed translation. Here is his rendering :—

—" Lo! the flickering wood above
And wavering shadow cast adown by darksome hang-
ing grove."

" Flickering wood " is not of subtle signifi-cance enough to suggest what is somehow con-veyed by the original phrase. I have seen the sunlight and a breeze playing at once through the bright-green top of a tall tree when the sudden thrills, so to speak, of golden fire, leap-

ing among the swaying foliage, were like
flashes of rare thought shot swiftly through
the brain of some grand genius.

Although I have hinted at the triolet, I shall
not speak of that, or indeed of any other
purely conventional form of verse, saving the
mere observation that nothing of the kind,
from the sonnet to the rondel, is suited to the
freshness and freedom of out-door life. The
over-racy honey of the bumble-bee, little suited
as it is to the table of the epicure, has such
flavor as ought to mark the songs of the sylvan
poet. I am in hopes that in our country a
school of young singers will soon appear,
widely different from that now forming in England, and also unlike the *jeune école* of France.
Why should we as a people foster, or even
countenance, forms of poetical affectation
worn out and flung aside by the Old World
some hundreds of years ago?

Our venerable Walt Whitman may have
pushed at times too far in the other direction,
but he has caught the spirit of freedom and
has dashed his unkempt songs with a dew as
American as that of Helicon was Greek. It is
a broad, out-door sense in which one enjoys
some of his breezy verses :—

"I think I have blown with you, O winds ;
 O waters, I have fingered every shore with you."

It is indeed a pleasing thing to idly blow
with the wind, or to blow with the wind for a
purpose ; and what is more recreating than to
finger sweet shores with the water ? A canoe,
if but a pirogue, and a shore to finger, if only
the bank of a rivulet, can give delight of no
uncertain sort to a healthy soul.

A Western poet, Ben Parker, has embodied
in a simple stanza a good idea of that freshness
which lingers in the memory after one has been
driven by the pressure of worldly cares out of
the redolent ways of nature :—

> "O morning when the days are long,
>   And youth and innocence are wed,
>   And every grove is full of song,
>   And every pathway void of dread ;
>   Who rightly sings its rightful praise,
>   Or rightly dreams it o'er again,
>   When cold and narrow are the days,
>   And shrunken all the hopes of men —
>   He shall re-waken with his song
>   The morning when the days were long."

The old English poet, Sir Richard Fanshawe,
took a gloomier view :—

> " Let us use it while we may
>   Snatch those joys that haste away !
>   Earth her winter coat may cast
>   And renew her beauty past :
>   But, our winter come in vain,
>   We solicit spring again ;
>   And when our furrows snow shall cover
>   Love may return, but never lover."

There was a philosopher for you ; but here
comes one of our young American poets with
a fancy that finds pretty and apt comparisons
wherever it skips.   Sings Edgar Fawcett :—

> " If trees are Nature's thoughts or dreams,
>   And witness how her great heart yearns,
>   Then she has only shown, it seems,
>   Her lightest fantasies in ferns."

It is quite surprising, when one comes to
look, how chary our later poets are of using
the dew for dampening their materials ; they
seem to prefer lamp-oil.   It may be, after all,

that lamp-oil is the better medium, but just
now I am writing from the saddle of a tricycle
with the spell of all out-doors upon me.

How precious is the pleasure now-a-days of
coming upon a really good stanza of verse, one
that breaks open, so to speak, like a fragrant
bud, and distils into one's mind the quintessence of genuine originality! I do not speak
of such originality as Poe's or Baudelaire's or
Rossetti's, but such as Swinburne has shown
in a choice few of his simpler lyrics, where he
has forgotten himself; for Swinburne is a
master when French and Greek influences do
not master him. His music is haunting, and
there are, scattered through his poems, pictures sketched from nature with a hand as free
and firm as Shakespeare's :—

> " Where tides of grass break into foam of flowers,
>   Or where the wind's feet shine along the sea."

It is not hard to find good out-door poetry if
we go back to the beginning of English verse.
Chaucer, with the language fresh in his hands,
so to speak, coined his phrases with a pen
dipped in dew. See how he begins his pro-
logue :—

> " When that Aprille with his schowres swoote
>   The drought of Marche hath perced to the roote,
> And bathed every veyne in swich licour,
>   Of which vertue engendred is the flour."

From Chaucer's day down to this no poet,
save Chaucer himself, has written four lines so
full of the subtle flavor of Spring as these. I
must add another stanza :—

> " And the river that I sat upon,
>   It made such a noise as it ron,

> Accordant with the birdes armony,
> Methought it was the best melody
> That might ben yheard of any mon."

Indeed, Chaucer is one of the few poets who are good companions in the open air. It is like a luncheon of fruit and nuts and choice old wine—reading the " Canterbury Tales " under a plane-tree by the brookside.

> " And he himself as swete as is the roote
> Of lokorys, or eny cetewale."

> —" Sweete as bragat is or meth,
> Or hoord of apples layd in hay or heth."

> " The hoote somer had maad his hew al brown,
> And certainly he was a good felawe."

Chaucer saw nature with frank, wide-open eyes, albeit he never forgot to be a scholar, as the times went.

> " And in a launde, upon a hill of floures,
> Was set this noble goddesse Nature,
> Of branches were her halles and her boures,
> Ywrought, after her craft and her mesure."

" To do Nature honour and pleasaunce " was so good, in the eyes of the old poet, that he did not nicely weigh the manner of the doing, viewed from the stand-point of our latter-day versifiers, but he let in the crispness of morning and the pungency of spring buds in lieu of these refinements of versification, now so highly prized. His knightly spirit and his courtly instincts could not repress his abounding love for the singing-birds, the breezy fields, and the wayside brooks. He was artist enough to know the value of words and the suggestive force of the more elusive elements of nature :—

> " Verse, a breeze 'mid blossoms straying,"

as Coleridge expresses it, was Chaucer's verse
in a large degree.   His was a *paradis parfumé*,
of a kind quite different from the hot-house
paradise of our modern poetry, whose odors are
of *l'huile de coco, du musc et du goudron* so liked
by Baudelaire and his admirers.

Emerson's poems are good to have in one's
tricycle-pouch.   I wish I could say as much for
those of Matthew Arnold. ˙Nothing can be
finer than the tonic raw sweetness of some of
Emerson's verses when read in the solitude of
the woods ; and no doubt this unstrained
American honey is too rich (as is the pulp of
our papaws) for the over-delicate English pal-
ate.   I am afraid that Mr. Arnold would find
fault even with the flavor of sassafras tea or
rhubarb pies !   It is one of Emerson's qualities, sharply observable, that, whatever may be
his technical short-coming, his thoughts are so
phrased in his poems as to give them a smack
of the clean, the home-brewed, the genuine.
A cup of sweet-apple cider, with its honest bouquet and non-intoxicating effect, is not a whit
more grateful than some of his wood-notes.
He had the nerve to preserve the aroma of a
thought, even at the expense of a false rhyme
or a halting verse.   He left some seeds and
floating bits of apple-rind in his cider.   As we
slowly imbibe his precious meanings we are
ready to quote him :—

> " I, drinking this,
> Shall hear far Chaos talk with me ; "   .

and we fall into a state of mind that melts

> " Solid nature to a dream."

Let some flying tourist stop for a moment

on a breezy hill-top, as I did lately, and read this :—

> " I hung my verses in the wind ;
> Time and tide their faults may find :
> All were winnowed through and through ;
> Five lines lasted sound and true."

Or this :—

> —" The bell of beetle and of bee
> Knell their melodious memory ; "

and he will feel a new consciousness of how Nature

> " Rounds with rhyme her every rune."

Scattered all through Emerson's poems are thoughts that cut into nature and tap her sweetest and most hidden veins.

It is remarkable that no Southern poet has arisen to give us the wood-notes of the land of the magnolia and the orange. Some of Sydney Lanier's verses, it is true, are dashed with the fervid colors of the semi-tropic, but he did not live to do his best, and his ill-health no doubt interfered with his out-door studies. His *Marsh Hymns* are lofty, fragmentary nature-songs, and I have no doubt that when his poems appear in book-form, as they soon will, it will be seen that his death was a sad thing for those who like genuine poetry. Still the fact remains that we have no poet who gives us the warm, odorous, fruitful South in rhythm and rhyme slumbrous as her sunshine and electrifying as her breezes. Indeed, no poet, of whatever country, has ever found the way to an expression of tropical out-door life. Of course I do not speak of mere descriptive verse, which is the lowest order of poetry. A

5

Southern Emerson would not be content with
mere adjectives of color and form ; he would
go about like a bumble-bee, extracting from
nature such sweets as might be found racy of
the soil. He would be a mole among the
juicy roots of plants, a butterfly among the
flowers. He would cut into the sap-veins of
the trees; he would peel the fragrant barks.
His poems would not be composed of these
things, nor principally of them, but their flavor
would come out of them, and out of the sunshine and the lazy summer winds.

Who knows but that the invention of the
wheel, this charming instrument of self-propulsion, is to work a new element into our literature—not merely the wheel element, but the
provincial element —an element which seems
to have almost disappeared from the poetry
and fiction made in the great literary centres
of New York, London and Paris. I have felt,
while enjoying short leisurely tours on the tricycle, that all the bright young cyclists of our
country are certainly in the best way of gathering that knowledge which fully complements
the lore of the books. Surely it is given to
him who knows Nature and loves her, to
speak :

> " As if by secret sight he knew
> Where, in far fields, the orchis grew."

IV.

Here are my notes of a short tricycle run
made on the second of May, 1884. The trip
was far more pleasing to me, no doubt, than
I can make it appear to others, but the notes
may serve to show how much can be seen,

heard, and felt in a little while under the ordinary circumstances of a run; or rather what a mass of observations one can record by the industrious use of one's eyes, ears, and note-book, and pencil, even when nothing really unusual occurs.

I set out quite early in the morning over a good road. A slight rain had fallen the day before, and there were a few puddles here and there, but no real mud. The spring had been a little slow coming, though the wheat-fields were waving ankle-high with a rich sward, and the woods were washed over with the tender green of tassels and leaves. A bracing freshness pervaded the air, which was from the south—a mere breath with a hint of summer warmth in it. No sooner had I cleared the town and got rid of the half-dozen ragged urchins that ran howling after me, as if I might have been mistaken for the advance agent of a circus, than I put on a spurt of power, bowling along in a level lane, with a hedge of *bois d'arc* on one hand and a high board fence on the other. A man walking in the middle of the road ahead of me evidently did not hear me coming, for when I whisked past him he shied like a young colt and glared at me as if he meant to attack me, but I left him so suddenly that I could not analyze his expression further. Somehow this little incident called up De Quincey's *Vision of Sudden Death*—a story which has always seemed to me a most perfect piece of art-work. If you have not read it, I advise you to take it with you on your first outing. It will fill an hour of rest with an enjoyment wholly

new. You will understand how it was recalled
by the trifling incident above recorded.

My way lay due east for nearly a mile, with
the meadow-larks whistling in the fields on my
right, and the woodpeckers chattering on the
fence-posts to the left. The woodpeckers
(those fellows half white and half black and
hooded in scarlet) had just arrived from the
South, and appeared overjoyed with their surroundings. They looked very clean in their
shining jet coats and snow under-garments.
A toll-gate stood at the end of the lane. I
whirled noiselessly through it before the woman who kept it could decide whether my
vehicle was down on her list, and ran over a
little hill just as the sun cleared the tree-tops
in the east. A small boy was riding a big-
wheeled plough, to which three fine sleek
horses were working abreast. The musty
odor of the fresh-turned soil was very pleasant.
Blue-birds were dropping into the new furrow
behind the plough to get the larvæ of various
insects exposed there. Two sparrow-hawks
were wheeling in small circles, some fifty feet
high, watching for field-mice, or possibly intent
on taking one of the blue-birds unaware.
There was a worm-fence on one side of the
road and the corners were literally carpeted
with wild blue-violets. What a pity it is that
these beautiful flowers have no perfume ! The
lack seems to take a great deal from their
value when one discovers it. It is almost like
finding that a very musical song has no meaning in its sonorous phrases. I now had some
stiff work going up a hill on a curve, and then
came a smooth bit of coasting, followed by a
short stretch through level heavy sand ; then

across a brook on an iron bridge and into a
grove of buckeye trees heavy with young leaves
and clustered blooms, about which the wild
bees were booming merrily enough.

Here I stopped, and sitting in the saddle,
sketched in the rough outlines of a boy who
was trying to snare sucker-fish, in a clear eddy
of the brook, with a looped wire. The first
Baltimore oriole of the season was singing
overhead in its peculiar, monotonous way.
This bird's song always seems spiral to me, as
if it had got a twist in coming forth. On the
anchor-posts of an old water-gate, I saw some
of the finest lichens I have ever met with ;
great round rosettes, puffed and ruffled, showing many delicate shades of sap-green, celadon
and gray. Not far from here I found a hill
too steep for comfortable riding, and after
pushing my machine up it, I was glad to see
before me a long stretch of level road through
beautiful farms. An apple orchard, too
closely set, was beginning to bloom, and a long
row of cherry-trees was white as a windrow of
snow. What is more expressive of comfortable, worthy wealth and liberal security from the
failures of life than a broad, well-kept Western
farm ? Here were fields of wheat, so wide that
they looked almost like prairies, side by side
with meadow-lands on which the clover and
timothy were thick and green over hundreds
of acres ; and then the rich black plough-land,
too, where soon the corn-planting would begin.
Orchards, garden-plats, grazing lands, cattle,
swine, sheep, and horses, broad-winged barns,
windmills for pumping water, and a spacious
residence embowered in maple trees ; surely it
is well to be an Indiana farmer.

I bowled along at a good rate with my head high, taking in deep draughts of the wholesome air; a long row of beehives in a garden, with the busy workers stirring on their little porches, sweetened the scene with a thought of big white honey-combs and snowy muffins. A fair, yellow-haired child was standing on a stile as I ran past the house, and she looked at me with great surprised blue eyes, holding meantime her little sun-bonnet in her hands. A big brown dog left her side and ran barking after me in a good-natured way for some distance, then turned and leisurely trotted back. A little farther on I stopped to watch a pair of cat-birds in a bit of hedge. They seemed to be looking for a good place in which to build their nest, for the female had a slender wisp of dry grass in her mouth. Up and down and in and out they went, all the time uttering their peculiar mewing cry. Finally the male mounted to the highest branch of the hedge and poured forth a sweet, trickling medley, not unlike the night-song of the Southern mocking-bird, though of far slenderer volume and inferior *timbre*. Why is it that the country folk have a contempt for the cat-bird? I have found this beautiful little songster under a ban from Michigan to Florida, with no one to say a good word for it, and yet, the mocking-bird and brown-thrush excepted, it has no rival in America as a singer.

Driving on again my road soon began to descend, growing steeper and steeper, until at length I put my feet on the rest, and, with hand on the brake, coasted at dizzying speed round a long curve down into a dense wood of maple, walnut, and plane trees that bor-

dered a little river. At the foot of the hill I
met a man driving a team of six horses hitched
to a wagon whereon was a saw-log sixteen
feet long and nearly four feet in diameter.
The log was tulip, usually called poplar in the
West, the *Liriodendron lulipifera* of the bota-
nists, and appeared not to have a blemish of
any sort in it. What a grand tree it must have
been when standing, and for how many Junes
it had bloomed in the woods, its huge flowers
flaming among its rich green leaves.

For some distance my road now skirted
the foot of a bluff along the bank of the river.
At one point I stopped for awhile to watch
a fisherman casting for bass. He was in a
little skiff near the middle of the river and
was casting down stream with a minnow for
bait. He appeared to understand his business, but I got tired, and drove on before he
caught anything; still I carried away with me
a pleasing impression,—in my memory a picture of the silver current breaking around the
skiff and the tall graceful angler patiently plying his rod and reel. What fascinating uncertainty there is in angling! What a big fish
one is always just on the point of catching!
As I write I have in my ears the murmur of
every brook from Canada to the chestnut-cov-
ered hills of North Georgia.

Turning aside from the main road I pushed
my tricycle up a steep, stony hill and mounting, soon found myself following the mean-
derings of a narrow cart-way, overshadowed by
wide-branching beech trees just beginning
to show their leaves. A half-mile of slow
riding brought me to a thicket of wild plum
bushes loaded with their fragrant white

blooms, amongst which bees and other insects were glancing and humming, and a number of small yellow green fly-catchers were actively engaged in a restless pursuit of their proper food.   I gathered a big bunch of these odorous plum-sprays and bound it fast to the handle of my brake lever so that I could have with me in my further journeying the fruity breath of the wild orchard.

Running down a long rut-furrowed slope, and then over a damp flat in a cool, shady hollow, I came to a nasty little stream sweeping through a narrow bog.   Here I called a halt for consultation.   That mud looked deep and treacherous.   I saw where a wagon had been pried out of it with fence rails.   There was nothing to do but get across, however, so I fell to work, carrying pieces of logs, rails, fallen boughs, etc., until I had made a quite respectable corduroy bridge, over which I pushed my machine with perfect safety; then I had to lift it over a large log that had fallen across the road.   In fact I did not mount again for a quarter of a mile, at the end of which I found myself at the source of the road, where it appeared that I was caught fast between a huge old red barn and a weather-beaten but comfortable looking farm-house.

A brawny, grizzled man with a hammer and monkey-wrench was tinkering with a disabled plough.   I approached him cap in hand, and mopping the perspiration from my face.   He immediately showed a deep but quasi-contemptuous interest in the mechanism of the tricycle.   I plied him with questions as to his crop-prospects, and was soon on easy terms with him.   I got a drink out of a sweet old gourd at his well, and obtained permission to ride across his wide pasture land to a road a half-mile distant. I mounted in his barnyard, and, while he held open a big gate for me, dashed out at my best speed into the level grass-field, where the dandelions shone like stars. A herd of steers, as I approached them, eyed me wildly for awhile, then ran away at a thundering pace, with their tails whirling and their heads high in air. I had to push across thirty acres of fresh ploughed land (a very uninteresting and tiresome operation) before I reached the road. Now came a long spin over a surface just damp enough to be elastic, and although the road-bed had been gravelled it was quite free from ugly stones. The air had freshened and was blowing in sweet gusts from the south.

The sunshine was growing in power; one could almost hear the buds exploding. A clover-field beside the road was a lovely sight, though not yet in bloom. Its dark green tufts looked as if they had gushed out of the earth in a moment of ecstatic impulse. Indeed, some occult force made itself manifest in every bud and blade, and stalk and leaflet, from which one could not fail to catch a fine mental tonic. I passed a level reach of maple wood in which grew scattered patches of mandrake that looked like the grass-green tents of lilliputian armies. In places the ground was rosy white with the blooms of the claytonia, or yellow with the stars of the adder-tongue.

What sweet and sure alchemic recipes Mother Earth gives us, if we could but read them! How unfailing are her schemes for the perpetuation of life, freshness, strength, beauty? These flowers are but the bubbles thrown up from her inexhaustible veins of vital force. Is not this woodsy fragrance which loads the air of spring mere surplus steam from Nature's alembics? and in breathing it do we not take into our blood a trace of her elixir? One's imagination renews itself by absorbing and assimilating the precious exhalations from the countless valves of woods and fields. How evenly and perfectly our book-lore blends and shades into what we gather from nature!

> " Spirit of lake, and sea, and river—
> Bear only perfumes and the scent
> Of healthy herbs to just men's fields."

All herbs and plants are healthy and wholesome, too, in their way. I saw a flicker eat the berries of the dreadful night-shade—not on this tour, for the plant comes later—and I have known a quail to swallow the seeds of the Jamestown weed with no bad result. But to my tricycling.

I soon came to where a broad road, leading homeward, crossed mine at nearly right-angles, and I set my face towards town with a three-mile run before me, over a fine rolling way between incomparably fertile farms. A fox-squirrel ran ahead of me on a fence until I came so near him that he sailed off into a field of wheat, and went bounding through the waving green blades to a lone walnut tree, up which he darted and disappeared in a hole.

The spires of our little city came in sight, gleaming above the maple trees that border the streets. I bumped across the railway track, whirled over a long hill, and descended into the suburbs with my blood tingling, and my memory full of fresh sights and sounds. At nine o'clock sharp I was at my desk.

THE THRESHOLD OF THE GODS.

" ———*Silva alta Jovis, lucusve Dianæ.*"

How shall we account for the old mythologies, or shall we attempt to account for them
at all? That beauty is imperishable, and that
whatever fills the measure of logic may be
taken as demonstrated, has somehow come to be
accepted by wise men as true. But shall we
receive or reject the gods of the ancients on
the score of beauty on the one hand or of logic
on the other? Who ever did believe in the
gods? Were they men of feeble minds or
debilitated physiques—a lot of degenerate
clods without any fixedness of character?
Was Agamemnon a fool, Homer a dunce,
Pythagoras a ninny, or Cæsar a weakling?

These may at first view seem questions both
trite and uninteresting; but I purpose sketching presently, as best I may, the outlines of a
quiet little adventure which led me to ponder
deeply over the proposition, Was there ever
any foundation in fact for this belief in the
gods? I will not say that I believe there ever
was, nor can I own to a total disregard for
certain rather obscure and mysterious evi-
dences in nature of the existence of beings
whose tenure of material bodies is as certain
and indestructible as the bodies are shadowy,
and whose power is somehow held, for some
reason hard to discover, in abeyance. If
gods ever were they now are. They may not
be now palpable or visible or audible, but

they are not dead or banished. For our present purpose let us admit that time was when
nature, the great generator of mysteries, dis-
closed immortal beings to man. Were these
beings necessarily, because immortal, omnipo-
tent or superhuman in their powers? I
should say they probably were possessed of
more than human potency in certain ways.
Immortality, even when robbed of everything
but the death-resisting principle, is in some
way very nearly married to invisibility in our
idea of it. The power of rendering itself invisible to human eyes, that is, the ability to
make itself a nonentity to all appearances, is
an attribute of every imaginable god, or at
least of every god at all like those of the Greek
and Latin mythologies.

Now suppose certain beings, born of a mysterious play of nature, possessed of these two
things, immortality and the power of rendering
themselves invisible, and what more is needed
as a basis upon which to build the fabric of
heathen polytheism? Why not, then, take the
so-called gods to have been a race of such
immortals, without any other attributes of the
true *Theos* in them? If such they were, how
natural for human imagination, operated upon
by the subtle influences of awe and wonder, to
add the rest. Indeed it seems to me hardly
fair, this laughing to scorn the beautiful theol-
ogy of the ancients without so much as giving
it the benefit of a charitable doubt, and without even admitting that it may have rested on
a venial mistake arising out of some manifestations of nature now withdrawn or in abeyance. But the gods may have been immaterial,
in the common sense, and yet not immortal in

the best meaning of the word. Certain conditions of mundane things might have been
necessary to their existence here. If we
should study nature closely for the purpose we
might discover those conditions.

And this fetches from its hiding-place my
theory. It may be called the grove theory.
No one can think of the gods as separable
from the woods and waters. The ancients ad-
mitted this. They went further, dedicating to
each deity its grove or stream. It seems to
me that this meant more than mere empty
complimentary dedication. It was a recogni-
tion and acknowledgment of the conditions
upon which the gods would remain with them.
In short, unsmitten, unshorn, pristine nature
could accommodate these mysterious beings,
and it only. The groves grown of virgin soil,
the uncultivated flowers and fruits, the balm
and spice of perfect trees—these prepared the
air for the gods to breathe. Something, we
may not know what—the keen pure essence
of unchanged nature from some source now
practically dried up, may be—fed them and
kept them within the bounds of visibility. The
dryads disappeared perforce, it may well be
assumed, when their woods were desecrated,
and the naiads when their fountains were pol-
luted. The fauns faded into shadows and
were blown away when the axe and saw had
felled the groves and fragrant thickets. The
satyr withdrew into the deepest recesses of the
forests as man advanced, and Apollo and
Diana fled away—whither ?

Possibly some secret potency existed in the
air that flowed through those virgin woods
and over those unpolluted streams which could

give to all those immortals the power of rendering themselves visible, and when it was
exhausted by man's encroachments they fell
away into invisibility. And if some hidden
cave in the world could now be found where
nature has never been disturbed by even the
simplest art, may be there might be discovered
one or two happy deities revelling in the merest pool, so to speak, of what was once the
great ocean of their " peculiar element." If
this theory is true the gods are invisible, not
dead, and they are invisible not from their
own choice, but because their " peculiar element " is exhausted which, while it lasted,
made visibility possible.

I have no certain recollection of having
been poring over this or any similar train of
semi-reasoning, nor have I the faintest knowledge of what I was thinking of, when my
guide, halting suddenly and knocking the
ashes from his pipe into the hollow of his
great brown hand, said, " Well, here we are."
At the sound of his rather gentle though deep
and sonorous voice, I looked around, feeling
as if I had been aroused from a dreamful slumber, without power to recall any definite idea
of my dreams.

Every one has experienced this feeling when
straying in an idle, musing way through some
still grove or quiet meadow. Suddenly, as if
by a spell of enchantment everything looks
strange. Even the sunlight is unlike itself.
The sough of the wind is peculiarly impressive.
Even the color of the grass is changed.
You rub your eyes; but it is some time before
you see, hear and feel natural.

So with me just then. I was well aware, to be

sure, that starting from the guide's cabin we had walked over a high ridge, almost a mountain, following for our way a zigzag path or trail that led us back and forth among vast fragments of variegated granite under wide-spreading boughs of low cedar trees. Now, however, we stood on the bank of a little river whose water crept past us in a slow but remarkably limpid tide as clear as glass, into which I gazed with an indistinct vision, and feeling a vague sense of the strangeness of everything about me. A pirogue lay moored at our feet. The guide motioned me to get in. I obeyed at once, but had time in so doing to note how old and frail, indeed how rotten the boat appeared to be. The guide accidentally tossed the pipe-ashes from his hand down upon one of the gunwales where they seemed naturally to disappear, mingling with the loose mould and minute fungi of the decaying wood. In this frail vessel we purposed passing over a dangerous rapid of the stream some distance below ; for it was the spirit of adventure had brought me here. I was in no condition, however, to realize the possibilities of the step I was about to take. I shook myself, rubbed my eyes and strove to get rid of this hazy mood; but succeeded only when the guide by a vigorous paddle-stroke sent us straight out to the stream's middle. Then I began to feel naturally and fell to making a close study of the guide and the boat.

What a taciturn, grimly selfish-looking fellow the man was ! His face was not a bad one, however, and his form was ease and strength incarnate. You could not guess such a man's age. Not a gray hair on his head,

not a wrinkle, denoting years, in his brow or
cheeks, and yet you suspected he was old. It
might have been the rather hard glitter of his
calm, gray eyes, or the half stolid way in which
he kept closed his immense hirsute lips, which
suggested something of senility coupled with
unusual strength. His bodily movements, too,
though full of elasticity of a certain sort,
lacked the ready suppleness of youth, suggesting instead the half-automatic, perfunctory
agility of long experience. You occasionally
see such old men by the sea or in the mountains. They are men whom age cannot con-
quer—the men of perfect health. But his boat
was not so impervious to time and exposure, it
seemed. A kind of dry rot had attacked it,
apparently years ago. This, however, seemed
to have added to its buoyancy, for it danced
upon the water like a bubble or a feather. I
could not help, as I glanced from man to boat,
imagining a sort of *rapport* between them, and
presently the odd fancy that, like the centaur
and the horse, they were really one, took hold
on my mind so forcibly that I could not re-
strain a low laugh as we began to glide down
the stream, so ludicrously did the blending of
the guide's gray, old clothes with the sides and
bottom of the gray, old boat, in color and texture, enforce the whimsical thought.

It may as well be stated here that the stream
upon which we were now afloat ran past the
guide's cabin over on the other side of the
ridge. But to do so it had to make a complete
double round a great point, after dashing
through a deep, hidden valley, down stony
precipices and between the close-drawn walls
of a resounding gorge.

My seat was forward near the prow of the boat, and I could look straight ahead over the little, decaying staff which, in imitation of a bowsprit, slanted off from the pirogue's beak. A glance down the river showed me how near to the dizzy escarpments of the mountain its current flowed, whilst over against this vast wall a wooded country, almost flat, swept off to a range of low green hills a mile distant.

The guide propelled our frail craft with a short, broad paddle which must have been very old, for the wood of which it was made had turned green and was curiously creased with worm-furrows and slimy with fungus or moss. Besides this paddle, a long cane rod, for use when the process of polling was necessary, lay at hand. But, so sensitive seemed our ancient pirogue to even the least impulse, there was little need of any engine, more than the stream's own current, to propel us withal. Noiselessly and evenly we slipped down the tide, much like the shadowy figures of a dream, it seemed to me, between the fern-braided banks. We scarcely made a ripple as we went. My habit of close observation soon prevailed over the dreamy mood that had settled upon me, and I began a minute study of the shores as they stole, by apparent motion, to the rear of us. Below the wild tangles of ferns and semi-fluviatic plants beautifully waved lines of parachrose stones lay in blending strata, as if half-welded by some process of fluxion long since ended—a dim polychrome rendered doubly effective by our motion. On the side opposite to the ridge the bank was quite low, giving us free insight to the farthest

6

glooms of the woods, where wild flowers of many kinds grew in profusion.

We had proceeded but a few hundred yards when I caught sight of a pretty, dappled fawn peering out at us with its great, mellow eyes from a clump of green shrubs. I now felt deeply vexed with myself for allowing the unreasonable importunities of the guide to cause me to leave my trusty rifle behind at his cabin. But a moment later, when the lissom, young animal against which I was aiming imaginary bullets sped away like the very spirit of merriness, I did not regret the gun. The common wild birds of the woods were everywhere. Blue jays and yellow finches, fly-catchers, nut-hatches and thrushes made a great chirping and twittering along with the mingled rustlings of their wings. I noted six or seven varieties of woodpecker, among them the ivory-bill and that great, scarlet-crested, black king of the woods named by the naturalists *Hylotomus pileatus*. Water fowls of the smaller kinds flew up before us, and occasionally a blue heron or a small wader of the bittern kind took wing in its peculiarly stately way.

A belted kingfisher, that most beautiful of all our birds of the streams, suddenly appeared in the air just in front of me, where he hovered for a moment as if doubtful whether to fly over us and go up the river or to turn about and retreat before us. He chose the latter. As he did so he uttered that sharp little laugh every angler has heard. O beautiful bird! your laugh has an evil ring! O *halcyon!* there is a great icicle in your heart, no matter how fine the weather you bring.

By short flights this bird kept a certain distance ahead of us, alighting now on a projecting stone of the cliff on one hand, and now on a reaching maple bough on the other, eyeing us warily as we approached and always laughing as it spread its gay pinions to float, rather than fly, down the steady little wind which drew along with the stream's course. We left all the other birds behind us. The herons and bitterns, describing the arc of a circle to avoid us, invariably turned up the stream in their flight, and the little sandpipers and shadowy looking waders of smaller kinds merely flitted from side to side of the water.

Sitting with my back to the guide and watching the halcyon's manœuvres, I began in an idle way to generate a fantastic theory connecting its flight with our own by a thread of fatalistic destiny. He, the beautiful, happy bird, was on the wind current; we on the water-stream. We were in a frail rotten canoe; he on his own splendid wings. How delightfully easy for him to evade death or even danger, whilst we, despite all exertions to the contrary, might soon speed right down to destruction! An underlying stone too near the surface could crush our craft into shreds. This bird of the hard, metallic laugh might be the demon of the stream leading us on to the rapids, to shout and scream and jeer when we were dashed to pieces in the cañon.

I noted now, by a glance, that our velocity was gradually increasing, and that we were following the sinuations of a sort of central current, which flowed among great bowlders and angular fragments of granite. The guide used the paddle merely as a rudder, and the

cane, only now and then, to push us away
from a dangerous breaker. The day, which
had been a singularly fine one, was now fast
drawing to its close, the sun having fallen behind the ridge, and a soft bloom hung directly
over us—a shadow overtopped by the vast
reaches of yellow sunshine. Our flight, however, would be short and the rapids would
swallow us, or happily we would swing round
the mountain's wall and slip down the gentler
current beyond to the guide's cabin, before the
coming of twilight, possibly before sunset.

The guide had described to me, in his
grimly laconic way, how he had frequently
passed these rapids for the mere excitement
of the adventure. I was the first man he had
ever led into this cove and he was sure that
no human being, himself excepted, had ever
before set foot here. This communication
was sufficient of itself to brace me beyond any
fear, even if I had been a most nervous man,
instead of a resolute naturalist used to danger.
Therefore I looked forward to the catastrophe
of this little drama with a calm mind and even
pulse, toying, meanwhile, with the curious
fancy that the halcyon was luring us on to destruction.

I was once talking with a great man, whose
profound knowledge and wise judgment would
seem to preclude trivial fancies from his mind,
and was surprised at hearing him tell how
often, in his moments of solitude, his imagination or fancy would fasten upon some insignificant thing as ominous or prophetic. A gay
beetle dancing in the sunlight before him; a
withered leaf blown across his path; a suddenly discovered violet or flower-de-luce; the

peculiar tone of a bird's voice; any, even the
least noteworthy thing, would hint to him of
the future. He would find himself trying his
fortune, so to speak, by little tests put in an
almost involuntary and wholly whimsical way,
to accidents and circumstances as they would
come of things as trivial as the mere breaking
of a twig or blowing away of a flower petal.
He related, with minute details, how once an
emerald-green, peculiarly brilliant scarabæus
kept itself by short, sudden flights, just ahead
of him in a woodland path, and how after he
had followed it some distance, wondering
what it was leading him to, he came upon a
huge rattle-snake, coiled ready for a spring.
The beetle had saved the life of a great states-
man and a true man !

I could not console myself with the fancy
that the kingfisher would steer us safely
through the rapids ; for his voice was insincere,
and his very movements would forcibly suggest
sinister things. Such is human perversity,
moreover, that I preferred the evil interpretation. I actually found myself gloating over
the anticipation of the halcyon's successful
stratagem. I even smiled as I saw, in fancy,
our boat dissolve into fibres and ourselves go
whirling through awful vortices mangled and
dead !

Nevertheless, I noted everything we passed,
and fixed in my memory with the power of a
trained concentration the changes in the
landscape bordering the stream. These
changes were constant, blending into each
other like colors on the artist's canvas. I imagined that the trees and shrubs and ferns,
and the aquatic grasses into which the mar

ginal ripples of the river leaped with low whis-
perings, constantly grew brighter and greener
as we advanced   Overhead the sky was
purely blue and clear, with just a hint of the
yellow sunlight flung athwart it.   In mid-air,
above the mountain's shadow, there hung a
misty splendor, such as is often seen on very
hot days hovering over water.   A fragrance,
which strengthened apace with our motion,
reached my sense, as if from some gradually
opened *pot pourri* of all sweet, spicy things.
The great, belted kingfisher seemed to feel
this as he led on, flinging back at us the chat-
ter of his voice and the rich, silken clash of
his wings.

I was now aware of an obscure feeling of
restless expectancy beginning to infuse itself
through me.   I turned half about to look at
my guide.   He made a frightful grimace at me
for rocking the boat, and glancing down I saw
some minute sprays of water bubble over the
gunwale !   Out through the momentary scowl
of the guide's face his vast age seemed to leer
like a wild demon.   Those bubbles leaping
over the boat's rotten side reminded me of
how easily it might swamp in the rapids.
With a little twinge of self-rebuke for my
thoughtlessness, I resumed my former posi-
tion.

Within these last few moments of time,
some change of no doubtful sort, but still a
change which eludes expression even now,
had taken place in the general appearance of
all surrounding things.   It may have been an
atmospheric or chromatic variance, it may
have been merely the mutations of the evening
shadows hovering in this low valley ; but, from

whatever cause, a something like the glamour of a dream or of romance had settled down upon stream and rocks and trees. An exhilaration like that induced by a salt breeze, more refined and subtile, however, took hold on me. The motion of the boat was now quite rapid, but smooth and noiseless.

I began to be impressed with the utter, the primeval, the unchanged beauty of the landscape. These woods, locked in by awful precipices, this stream, full of dangerous falls, had never been troubled by hunters or anglers, or naturalists or tourists, nor yet by the insatiable makers of farms. Pristine power and perfectness dwelt here as they did æons ago. I looked and saw the smooth, greenish-colored bark of the trees, the deep expression of *riant* vitality in the leaves; I drew into my gratified sense the strengthening bouquet of surrounding nature, and then suddenly the inquiry, from what source I cannot say, arose in my mind, are the gods still here? At first it was a half-idle thought, blown across my mental field like a rose petal across a garden; but it found a lodgment. I toyed with it and it grew. It suited my mood and the mood of nature.

The halcyon flitted on before us, and now, far away, like the soft murmur of a breeze, our ears caught the pulsating sound of the rapids, A deer, bearing young antlers, stood on the bank and very steadily eyed us as we passed. He did not seem to fear us, his gaze denoting only a lively curiosity. Indeed he had no cause to fear us, for all thought of the chase was far from me, and as for my guide he had enough to do caring for the boat.

Are the gods still here? The question fed my fancy. I began, in a half-earnest, half-idle way, to scrutinize every dim opening, every shadowy recess of the woods, as we sped by. I wove a cocoon of the old, silken webs of poesy around about me, looking through the sheeny film of which I hoped to assist the shy deities in taking on visibility. If I could only see one god, even though it flitted past me a ghostly, diaphanous mockery of its former self, what a joy it would be!

The wings of our luring halcyon were now in almost constant motion, so swift was our following, and the sound of the voice of the waterfall was deepening and spreading. Some little thrills of quietly ecstatic delight began to trouble my senses. I have occasionally felt the same when sailing before a smart breeze in an open boat after a long absence from the sea.

At some distance before us I saw a shining line drawn, like a wavering gossamer, across the surface of the river. Beyond it a silvery mist swayed in the gloom of giant trees that partially overshadowed the water. This line was the break where the cataract began and this mist was the spray from the agitated stream in the cañon; but to my mind the silvery thread was the index of something more, and with a leap, so to speak, my imagination reached the threshold of the gods! The line marked the boundary of the haunts of the shining ones. Heavy and sweet the odors drifted upon us, and in all the trees we heard a satin rustle. The cardinal-birds and the wood-thrushes suddenly ceased their singing. Deeper and deeper we sank into the narrowing

dell, sweeter and softer the gloom grew apace.
I marked well the giant trees just beyond the
sheeny line, and saw through the spaces between them shadowy mysteries flitting to and
fro—mysteries that a dash of sunlight would
have dissipated, that a puff of wind would have
lifted up and scattered like smoke. Faster
and faster we sped, wilder and wilder grew
the flight of the halcyon. He could not take
time now to light at all, but only to hover a
moment at eligible perching places, and then
hurry on before us.

What a thrill is dashed through a moment
of expectancy, a point of supreme suspense,
when by some time of preparation the source
of sensation is ready for a consummation—a
catastrophe! At such a time one's soul is
isolated so perfectly that it feels not the remotest influence from any other of all the uni-
verse. The moment preceding the old pa-
triarch's first glimpse of the Promised Land
—that point of time between uncertainty and
certainty, between pursuit and capture, where-
into is crowded all the hopes of a lifetime, as
when the brave old sailor from Genoa first
heard the man up in the rigging utter the
shout of discovery—the moment of awful hope,
like that when Napoleon watched the charge
of the Old Guard at Waterloo, is not to be
described. There is but one such crisis for
any man. It is the yes or the no of destiny.
It comes, he lives a life-time in its span; it
goes, and he never can pass that point again.

But there are crises, scarcely less absorbing, to which, after they are passed, one can
turn and almost live them over. These are
the crises into which no element of selfishness,

more than the mere modicum contained in the anticipation of pleasurable sensations, has entered, or crises of the imagination based wholly on phantasmal exigencies. I reach back the powers of my memory now, and they fetch up out of the past, even to the minutest detail, the whole of that little period of time during which I waited, with bated breath and condensed expectancy, to see a god!

The river was bearing us on at a rate of speed which, but for the silent evenness of the motion, would have been frightful under better circumstances. But the wood of which the pirogue was made—it must have been yellow tulip—seemed so unsound and semi-disintegrated that the wonder was it did not dissolve into a flake of vegetable mould upon the water, and thus let us sink!

A vast white bird, probably a snowy heron the *Garzetta candidissima* of our naturalists, swept majestically across from side to side of the river, directly over the mysterious shining line and just hitherward of the pale mist, quickly losing itself among the trees. Again I saw, or imagined, shadowy forms stealing through rifts in the flower-sprent glooms of the woods. But they were less satisfactory than the dimmest forms of a dream. I could not follow them a second of time.

A broad booming heralded our approach to the cataract. We felt no motion, so steady was our sweep, and yet we were leaving the dreamy wind behind us. Halcyon, with erect and dishevelled crest, led on in an ecstasy of chirp and flutter. I became aware, through some slight, ominously decisive movement of the guide, that he was preparing for a supreme effort. We were nearly opposite a grand opening in those stately trees, out of which seemed to issue the silvery line which cut the river. I leaned forward, with suspended breath, to catch a glimpse right down it as we should pass. The gods were there, I knew they were ; I should see some one of them, at least, if only a sylvan faun or satyr, or a dryad slowly withdrawing into the heart of a tree. *Deus ecce! Deus.*

That great white bird came out of the shad-ows of the woods again, and curving its flight down the stream seemed to melt into the mist. A sensation of dewy coolness crept over me, as if shaken from the rorid sandals of some passing naiad. The bank of the river opposite to the ridge's precipice now presented a gay, almost fantastic appearance. Tall, aquatic grasses, thinly interspersed with certain scar-let-spiked riparian weeds, were sown at the water's verge ; their long slender stalks and semi-translucent leaves, waving to the impulse of air and water-ripple, sent forth a sort of shimmer like that which Virgil intended to describe with the phrase " *Tum silvis scena coruscis* "—a waving motion with light flashing and flickering through. Right opposite this a narrow, vertical rent intersected the ridge, and through it an almost level finger of the sun reached to caress the grass. Just as we passed I noted, by an instantaneous glance, a strange and beautiful thing—a troop of dragon-flies, purple-bodied and silver-winged, filing rapidly, in open order of ones and twos, across the sunlight into the dewy recesses of the river's fringe. Each gaudy insect, as it flew, wavered in the air so dreamily and eccentrically that

somehow I was reminded by their course of
those shadowy, silvery lines in the blades of
Damascus daggers.   •

We slipped on and on, still following the
now madly careering halcyon. For the merest point of time, not long enough for an eye
to twinkle, we were opposite the rift in the
woods and trembling on the verge of mystery.
I looked down the open vista and saw something, I know not what—a form or a shadow,
an image conjured up by my imagination, or
only a blending of the glooms and gleams by
force of distance and velocity—but a new element was added to my nature. I felt a great
thrill. A new joy took root in my heart. A
new flower blew open in my soul. *Accipio
agnoscoque deos !*   `

It seemed that down that aisle I could look
to the remotest age of time ; and out of it,
blowing into my eager face, I felt the unchanged, the unchangeable spirit of Eld !
Was it, or not, a face that I saw ? Can I ever
know ? The flowing hair, like blown supple
ringlets of gold floss, the gray deep eyes, the
divinely smiling lips ; were they not there ?
And the shining body and agile limbs, did I
only fancy I saw them ? How shall I ever be
sure ? *O ! Dea certe.* An indescribable something, as of that whole landscape melting and
vanishing, by a sudden and noiseless deflagra-
tion, followed close upon this fortunate moment. With a harsh, maniacal cry of delight,
the belted halcyon leaped over the coruscating
line into the silvery mist beyond. And, like
an arrow flung from the bent bow of the river,
we were whirled after him into the vast fanged
jaws of the cañon.

I felt our pirogue leap and shiver; I heard awful noises, as of battles and storms and tumultuous applaudings, as of a million clapping hands, as we rushed down into the rent hill. *Sic, sic juvat ire sub umbras!* All above us the mist and spume boiled and rolled; all below us the mad waves leaped and fought; all round us the gray, wet fragments of granite offered destruction. Nature's wildest frenzy of passion was bearing us down, down, down! O, the calm madness that seized me! It was awe traced in marble—it was terror frozen in ice! O, the sweet vision, so suddenly mine, so abruptly gone! A mysterious joy, like the memory of a heavenly dream, lingered in my heart, down deeper than any fear of death could go.

Deeper and deeper we plunged down between the dank, fantastically grooved jaws of the gorge, till the mist and darkness blended into one, and the thunder of the stream in its agony was appalling. Even this did not drown the metallic laugh of the halcyon as it led on through that horrid tumult. I felt a wet wind rushing over me, I saw the spume sparkle like phosphor, whilst the shark-like teeth of the walls on either hand drew closer upon me.

How deep the ecstasy below us! How far-reaching the immitigable storm-mist above us! How old and worn the stolid stones about us! O threshold of the gods, what a distance behind us! O sweet, calm, every-day world, how infinitely removed from us!

Finally we felt a mighty swell lift us and savagely shake us. A heavy spray dashed over us, and our frail vessel quivered and quaked, as if in a convulsion of pain. Suddenly the gorge closed up till the slimy walls thereof oppressed us and its jagged teeth grazed us on either side. But on we rushed, tempest behind us, thunder before us, the blackness of utter darkness all about us, and at last, with a mighty explosion of all terrors, we were hurled like a missile from some giant engine—a very missile, indeed—forth from the grim, stony lips of that awful fissure, reeling and spinning far out upon the swift, level bosom of the little river lapsing into the open country.

The evening farewell of the sun was glorifying the distant mountain lines, the sweet maple trees on either side of us were waving betwixt gloom and splendor, and the breeze was a deep, tender sigh of relief.

> " *Unde hæc tam clara repente*
> *Tempestas ?* "

The belted halcyon turned aside in his flight, and perching upon a bough laughed his fill at us as we drew past him. The roar of the rapids receded and faded, leaving at last in my heart a tender melody which never can depart. I had hovered on the THRESHOLD OF THE GODS!

BROWSING AND NIBBLING.

I WAS once following a tireless guide through a wild mountain region of the South, when, in answer to a direct question, he delivered himself as follows:—

"What makes me allus a-nibblin' an' a-browsin' of the bushes an' things as I goes along? Well, I dunno, 'less hit's kase I've sorter tuck a notion to. A feller needs a heap o' nerve ef he 'spects to be much account for a deer-hunter in these here hills, an' I kinder b'lieve hit keeps a feller's heart stiddy an' his blood pure for to nibble an' browse kinder like a deer does. You know a deer is allus strong an' active, an' hit is everlastin'ly a-nibblin' an' a-browsin'. Ef hit's good for the annymel hit orter be good for the feller."

This philosophy immediately gained a lodgment in my mind. I delightedly took up the seeds of suggestion let fall by the strong-limbed, steady-nerved mountaineer, and forced them to rapid quickening and utmost growth. The old alchemists in their search for the elixir of life ought to have known that the birds and the animals of the wild woods had long ago discovered it. How many sick deer, or bears, or partridges, have ever been found by hunters or woodsmen? For twenty years, as boy and man, I have been an untiring and persistent roamer in the wildest nooks and corners of our American forests; and during this period, I have never found a deer, a bear, a squirrel, a turkey, a grouse, a quail, or any

wild bird, suffering from any fatal ailment
other than wounds. When their food is
plentiful all kinds of wild things thrive. Of
course, when unusually hard winters come,
and food cannot be found, the non-migratory
birds and animals suffer, often to death, from
hunger and cold. But this is accident rather
than anything else. Take a healthy child into
the woods, and see how naturally and surely it
will fall to nibbling at the buds, and bark, and
roots of things. There seems to be an innate
hunger for this sort of food, lying dormant in
every human being until called into activity by
some association, accident, or exigency.

Now, I am not going into the dear old
theory of the botanical doctors touching nature's remedies for man's ailments. I am not
a physician, and I favor no special school of
medicine. But I do maintain that it is good
for man—and woman, too—to nibble and
browse. Go bite the bud of the spice-wood,
or the bark of the sassafras, and tell me
whether you feel a new element slip into your
nature. No sooner do you taste for the first
time this wild, racy flavor, than you recognize
its perfect adaptation to a need of your life.
Nor is this need a mere physical one. Somehow the fragrance and flavor that satisfy it
reach the thought-generating part of one, and
tinge one's imagination and fancy with new
colors.

I remember, with a steady delight, some days
spent with the ginseng-diggers of North Carolina. It was there that I first tasted this
celebrated American root, and discovered a liking for its charming, aromatic bitter-sweetness.
No wonder the Chinese prized it above gold!

These ginseng-diggers—or "sang-diggers,"
as they are called—are queer folk ; very interesting in a way, ignorant, superstitious, strong,
stingy, and honest—a sort of mountain tribe
to themselves. I followed a company of them
around the jutting cliffs and fertile " benches "
of the Carolina mountain region, until I really
had grown to like their careless, nomadic life,
with its flavor of chestnuts and ginseng. In
the spring is the time for browsing; in the
autumn comes the nibbling season. The
squirrels begin eating the buds of the hickory
trees so soon as the sap has risen into them
sufficiently to make them swell. Your knowing squirrel-hunter cleans up his rifle about
this time, and visits every hickory tree in his
neighborhood. Somewhat later the grand
tulip trees begin blooming, and then the squirrels transfer their attention to them. A few
weeks of browsing in the spring woods will
make one acquainted with the characteristic
taste and fragrance of almost every tree, shrub,
and plant of the region.

True, there are a few—very few indeed—
poisonous things, and these must be avoided.
Nature has her evil streaks, running at wide
intervals through her opulence of good ; but
they are easily discoverable. Who would
ever be so obtuse to danger as to nibble at the
buds of the poison ivy ? This browsing-time
is also the season of our sweetest and most
charming flowers. While one is biting through
pungent barks and aromatic buds, one also
gets the benefit of perfumes as wild and witch-
ing as are the blooms from which they exhale.
I do not know how to explain the influence of
the bitters and sweets, the acids and sub-acids,

7

the aromas and perfumes of wild things; nor am I sure that explanation would be profitable, if possible. To taste the perfectly distilled honey that lurks in the red-clover bloom is a sufficient demonstration of this influence. A subtle thrill, elusive as it is fascinating, follows the touch of the tongue to this infinitesimal philter. It was made for the bumblebee; but your pastoral man may profit by the insect's example. If Rossetti, while bending over a woodspurge, had been less an artist and more a poet and philosopher, he might have discovered more than he expresses in :—

> " One thing then learnt remains to me,—
> The woodspurge has a cup of three."

Compare the flowers of Tennyson and Keats with those of Baudelaire—

> "Des fleurs se pâment dans un coin"—

and the whole fearful difference between the sweets of nature and the filth and rottenness where those sweets are wanting, will rush upon your consciousness. There is something more than the mere shimmer of rhetoric in Virgil's

> " Tum silvis scena coruscis
> Desuper, horrentique atrum nemus imminet umbrâ."

There is in the words a suggestion of what woodsy freshness and fragrance, of what spices and resins, that grove may hold. Howells brings to mind the same possibilities when, in his poem called " Vagary," he sings—

> "Deep in my heart the vision is,
> Of meadow grass and meadow trees
> Blown silver in the summer breeze."

There is a smack of browsing in such a verse as—

"But in my heart I feel the life of the wood and the meadow."

And when Keats forgets the Greek myths and turns to pastoral memories, how true and fresh and fine his note ;—

"I cannot see what flowers are at my feet,
    Nor what soft incense hangs upon the boughs;
But, in embalmed darkness, guess each sweet
    Wherewith the seasonable month endows
The grass, the thicket, the fruit-tree wild."

But we poor clay mortals, who have never been able to get within the charmed life of the poets, can have our sip of honey-dew, and our morsel of wild balsamic resin, our mouthful of pungent buds, and our taste of aromatic roots, notwithstanding our coarse natures, just as well as these successors of the gods. Still, I fancy that it is the literary man and the artist who get the most out of our out-door browsing and nibbling. Wild plums and haws and berries, papaws, nuts, grapes, and all the fruits of ungardened nature, have something in them to feed originality. One cannot chew a bit of slippery-elm bark without acknowledging the racy charm of nature at first hand. Children like all these things, because their tastes are pure and natural. Poets like them, because poets are grown-up children. Painters like them, because painters affect to interpret poetry and nature. Clods, like you and me, reader, like them, because they are racy and good; because they take out of our mouths the taste of artificial food, and because they seem to strengthen our connection with untrimmed and uncultured nature. They are, in their way of laying hold on our taste, like the poetic myths of the Greeks. They cloy for a

time, but when their season comes round again
the zest comes too

Was it not Adonis, as Shakespeare has it, to
whom the birds—

" Would bring mulberries, and ripe red cherries "?

To me the flavor of our American wild cherries has always been especially alluring. So,
too, the service-berries, with their wild red
wine, have tempted me to many a dangerous
feat of climbing. Often in the dense huckle-
berry swamps of the South I have refused to
be frightened from my purple feast even by
the keen whir of the rattlesnake's tail, though
the deadly sound would make my faithful dog
desert me in cowardly haste.

Along the banks of the streams of Georgia
and South Carolina grows a grape, known by
the musical name of muscadine, which I esteem
as altogether the wildest and raciest of all
wild fruit. Its juice has the musty taste of
old wine along with a strange aromatic quality
peculiarly its own. On splendid moonlight
nights I have swung in the muscadine vines,
slowly feasting on the great purple globes,
while the raccoons fought savagely in the trees
hard by, and a clear river gently murmured
below. Next to the muscadine among wild
fruits I rate the papaw as best. It is genuinely wild, rich, racy, and, to me, palatable
and digestible. I once sent a box of papaws
to a great Boston author, whose friendship I
chanced to possess, and was much disap-
pointed to learn that the musty odor of the
fruit was very distasteful to him. He fancied
that the papaws were rotten ! I dare say he
never tasted them ; and if he had, their flavor

would have been too rank and savage for his
endurance.   :

The gums and resins of our woods are few.
The sweet-gum, or liquid amber, is the only
genuinely fine morsel of the sort to be found
within the boundaries of the United States.
It is a clear amber fluid (flowing from any cut
or wound in the tree), which soon hardens into
a stiff, translucent yellow wax, possessing a
pleasing aromatic taste and odor, strangely
fascinating. One does not care to eat it; but,
once a lump of it goes into one's mouth, one
chews it until one's jaws are tired. I remember, when I was a very little child, going to a
backwoods school in Missouri, where all the
pupils, both great and small, would chew
liquid amber from morning till night; the
teacher chewed tobacco.

Browsing and nibbling has led me to taste
the inner bark of ·nearly every kind of tree
growing in American woods. The hickory
tree has a sap almost as sweet as that of the
maple, but it mingles with the sweet a pungency and a slightly acrid element of taste at
once pleasing and repellent to the pampered
tongue. The oaks have much tannin in their
bark, the astringency of which draws one's
lips like green persimmons; but the very
innermost part, next the wood, is slightly
mucilaginous and faintly sweet. Speaking of
persimmons—after a few sharp frosts this
wild fruit becomes mellow and rich, but to the
last retains a certain drawing quality, a trace
of that astringency already mentioned, which
keeps it from being a favorite, save with the
opossums.

There is no other woodland influence, however, so strong and fine as the perfumes, odors, and aromas. Of these each season has its own—the perfume of spring flowers, the odors of summer mosses and sweet punk, the aroma of buds and barks and gums. Even in midwinter, when a warm time comes, and the snow melts, and the ground is thoroughly thawed, there are woodsy odors borne about by the drowsy winds. In fact, the fragrance of January is sweeter and more subtly elusive than that of May. Go nibble the brown, pointed buds of the beech tree in midwinter, and you will find how well the individuality of the trees is condensed in those laminated little spikes. You taste the perfume of tassels and the fragrance of young leaves; there is an aromatic hint of coming nuts. You may almost taste the songs of the spring birds! What words these buds are! How prophetic! We bite them, and, lo! the spring rises in a vision! Its poem is read in advance.

I recollect a clear fountain of cold water around which grew festoons of cress and mint. I had been chasing the wild things all the morning, as a true huntsman will, and now I was tired and thirsty. At such a time what could be more welcome than mint and water? How soothing the fragrant flavor and the cooling draught! Then came the biting spiciness of the cress, to reinvigorate my nerve withal. Out of my pouch I drew a cake of maple sugar, and feasted like a god.

When winter begins to come on, the nuts come too. I cannot understand the taste of those who do not like the rich oily kernels of the butternut, the hickory nut, and the sweet acorns of the pine oak. Squirrels know

which side of a nut is buttered. They have
long ago learned that it is the inside. From
Florida to Michigan one may run the gamut of
nuts, beginning with the lily-nuts, or water
chinquepins, and running up to the great
black-walnut, including every shade of flavor
and fatness. They are all good. They were
made to eat in the open air ; and he who takes
them, as the squirrels do, after vigorous exercise in the woods, will find great comfort in
them. I cannot rank the artist or poet very
high whose stomach is too aristocratic for
wild berries, nuts, and aromatic bark. I fear
that such an one has long since allowed
that trace of savage vigor, which made him
of kin to Pan and Apollo, to slip away and be
lost. Shall we doubt that Burns got his sweet
strength and freshness, in a great measure,
out of the cool, fragrant loam his ploughshare
turned ? The gracious ways of nature are so
simple and so manifold. She gives up to us
by such subtle vehicles of conveyance the
precious essences of suggestion. She draws
us back from overculture to renew our virility
with her simples. She gives us dew instead
of philosophy, perfumes instead of science,
flowers in place of art, fruit in lieu of lectures,
and nuts instead of sermons.

In the manifest life of an individual no element is so pleasing as that trace of force
which suggests his kinship to wild nature.
Out of this springs a sweet stream of originality
and freshness, a sincerity and outrightness of
thought and action, of great value *per se.* I
have met men whose talk was spicy and aromatic ; from whose lips simple words fell with a
new, racy meaning. Their thoughts were red-

olent of the odors and essences of buds and
flowers, and sweet, mossy solitudes. Theirs
had been the oil of nuts instead of the oil of
the lamp.

There is no safety in culture if it leads to
artificiality. There must be a safety-valve to
any high-pressure system, social, moral, or intellectual. The connection with the sources
of nature must be kept perfect. Poetry,
painting, sculpture, and all the cognate elements of high education and sweet intellectual
attainment, must become mere manifestations
of a diseased fancy and imagination whenever
this connection shall be permanently severed.
It matters little by what slender streams nature feeds us, so that we get the food at first
hand. History seems to teach us that utter
artificiality is the forerunner of decadence. On
the other hand, in the flowering time of a people's youth come their geniuses. England
can have no Shakespeare now, Germany no
Goethe, Italy no Dante. Culture has gone
too far. The wires are down between nature
and the leaders of fashion in fine art. True,
we have the microscope in the hands of hundreds of analysts and fact-gatherers; but this
serves only the turn of the men who despise
every element of nature that cannot be con-
trolled for the furtherance of the demands of
artificial life.

Reader, let us go out occasionally to browse
and nibble, and gather the savage sweets of
primeval things; to revel in the crude materials of creation; to get the essential oils, the
spices, the fragrance, the pungent elements
of originality.

OUT-DOOR INFLUENCES IN LITERATURE.

THE earth is the great reservoir of physical forces, and whilst no scientist has yet been able to discover how intimate or how perfect is the connection between the mental and the physical, there exists, no doubt, a correlation between the processes by which the body and the soul are kept healthy and vigorous by draughts on the great reserves of Nature. One grows tired of books and cloyed with all manner of art. Then comes a hunger and a thirst for Nature. Real thought-gathering is like berry-gathering—one must go to the wild vines for the racy-flavored fruit. Art and Nature are really the antipodes of each other— one is original, the other second-hand. When we go from the library or the studio to the woods and fields, we go to get back what Art has robbed us of—the freshness of Nature. Art presents compositions; Nature offers the original elements. The suggestions of Nature come, as the flowers and leaves and breezes come—out of the mysterious, invisible generator; but Art merely reflects its suggestions back upon Nature.

What genuine poet or novelist has not caught his charmingest conceits from some subtle and indescribable influence of out-door things? In-door poets, like Dante G. Rossetti, always lack the dewy freshness of Helicon, the thymy fragrance of Hybla, no matter how much of

the true maker's *labor limæ* may appear in
their works.   Even Poe and Hawthorne disclose too heavy a trace of the must and mould
of the closet.   Each stands alone, inimitable,
in his field, but lacking that balmy, odorous
freshness of the morning woods and pastures,
when the convolvulus and the violet are in
bloom.   We should have little faith in the
bird-song described by either one of those
wizards of romance.

> " The skies they were ashen and sober,
>   The leaves they were crisped and sere,"

in all their works.   Cheerfulness and enthusiasm have always seemed to me to belong of
right to the best genius.   Shakespeare exempli-
fies it ; the sublime audacity of Napoleon I.
instances it.   But Shakespeare was a poacher,
and Napoleon loved to dwell out of doors.   I
hold that communion with Nature generates
lofty ideas, feeds noble ambitions.   The only
way to lengthen a yard-measure is to gauge
each new length of cloth by the preceding one,
and not by the yardstick.   The growth will be
slow, but amazingly sure.   So in Art, if we
cast aside the standards and permit such ac-
cretion as Nature suggests.

But there must be some excuse for going
out alone with Nature other than the avowed
purpose of filching her secrets and accumulat-
ing her suggestions ; for, as a matter of fact,
nearly or quite all of the available literary or
artistic materials caught from her great reser-
voirs come without the asking, and at the
moment when they are least expected.   Then,
too, the human mind seems to have no voluntary receptivity.   The power of taking in new

elements seems most active in the brain when the pleasurable excitement of a rational pastime is upon it. The artist is often surprised, while aimlessly sketching in the presence of Nature ; at the sudden coming on of a genuine "inspiration"—a suggestion leaping out of some accidental touch, or out of some elusive, shadowy change in the phases of things.

The direct study of Nature is dry, and its results, however useful and entertaining, far from satisfactory from a literary or artistic standpoint. As one can see an object better in the night by not looking straight at it, so the indirect view of Nature is best for the discovery of those inspiring morsels upon which the gods used to feed, and with which the poet, the novelist, and the painter of to-day delight to stimulate themselves. But the gods were hunters and athletes, as well as lyrists and songsters. They bent the bow with as much ease and delight as they blew in the hollow reed or thrummed on the stringed shell. They robbed the wild bees of their honey, and chased the deer over the hills ; they followed the streams of Arcadia, and haunted the fountains and glens of both Italy and Greece. The poets are said to be the successors of the gods. The gums and resins, the spices and saps, the perfumes and subtle essences of Nature make their nectar and ambrosia. It is the presence of this flavor of Nature that discloses the work of a genuine genius. No amount of cunning artisanship can create, it can only build. Genius works with animate materials and essences ; its

"Conscious stones to beauty grow."

In a bit of verse I once tried to express my idea of the true poet :—

> " He is a poet strong and true
> Who loves wild thyme and honey-dew,
> Who, like a brown bee, works and sings,
> With morning freshness on his wings,
> And a gold burden on his thighs,
> The pollen-dust of centuries."

This pollen-dust is to be found in the old woods as well as in the old books. The flowers of poesy are but impressionist sketches of the flowers of Nature. The little bloom of the partridge-berry has sweeter perfume than any lyric of Theocritus or Horace. From the proper point of view the big, vigorous flower of the tulip-tree is as full of racy, unused suggestions as it is of stamens. Virgil and Tennyson, Theocritus and Emerson, Sappho and Keats, have filled their songs with the most delicately elusive elements of Nature caught from out-door life. They are the half-dozen poets of the world who have come near in their work to the methods of the bee. The honey-cell and the poem are of divine art—the honey and the idea of the poem are of divine nature. Rossetti and Poe builded lovely cells, but they had no wild-flower honey with which to fill them; theirs was a marvellous nectar, but it was gathered from books and art. " Volumes of forgotten lore " served them, instead of brooks, and fields, and woods, and birds, and flowers.

Now, literature is not the whole of life, nor is the study of Nature the whole secret of literary inspiration. But recreation of body and mind is drawn from obscure and various sources, and the well-rounded genius seems to feed itself upon Nature much more than upon books. A book is most useful as a literary helper, when it may be used as a glass with which to better view Nature. I would not be understood as saying that all worthy literature is or should be a mere interpretation of out-door life ; far from it. Out-door life, I may say, furnishes the inspiration, the enthusiasm, the freshness. It furnishes the water for the clay, it gives the hand its certainty, the mind its new leases upon youth. It does not make the mind nor the hand ; it merely informs them with the creative effluence of Nature, as Thoreau would express it. It has a fertilizing power—this lonely communion with the out-door forms of life—which one may trace in the best works of the geniuses of all ages. Pan, when he pursued the flying Syrinx, and at last clasped an armful of reeds instead of the nymph, very accurately typified the poet. He took the reeds and made of them his pipe. He had caught the idea of music from the sounds of the rustling leaves and stems. If you would like to fully understand the meaning of this myth of Pan and Syrinx go clasp an armful of wild green reeds and hold your ear close to them. You will hear the sound of washing seas and rippling rivers and flowing breezes all blending together ; voices from vast distances and snatches of immemorial song will come to you. Like Pan you will long for a pipe, that you may express what has been suggested to you by the reeds.

Awhile ago I said that direct, conscious study of Nature was not best for gathering those impressions most valuable to the poet and artist. Thoreau is a striking example of a poet spoiled by this direct study. Compare his poetry with that of Keats or Tennyson or Emerson, and it will be discovered that his obvious attitudinizing before Nature prevents him from appearing sincere, simple, and fresh in his conceits. It seems that the available material which one gets from Nature, save for scientific purposes, must be received aslant, so to speak—must be discovered by indirect vision—and while one is looking for something else. Thus while Thoreau was besieging Nature for her poetic essences, he failed to find them, though Keats had stumbled upon them apparently by accident.

> " What melodies are these?
> They sound as through the whispering of trees."

If ever the songs of a poet

> "Come as through bubbling honey,"

and

> "In trammels of perverse deliciousness,"

the songs of Keats did, and in them we may find in the best measure the influences of the indirect study of Nature.

Now, there are few persons who, like Keats, will absorb these influences without some stimulus other than the poet's love of solitude; nor is solitude for its own sake wholesome. On the contrary, it is inimical to healthy physical and mental development. Keats' might have lived to finish all his " divine fragments " if he had been an enthusiastic canoeist, archer, or bicyclist. He died of consumption at the age of twenty-five years! If William Cullen

Bryant had possessed Keats's genius, of if Keats had had Bryant's physique! Think of the boy-author of *Endymion* singing till he was eighty! And yet such a thing might be if recreation were regular and judicious. If Keats were alive to-day he would not be ninety· years old, and yet his poems have been classics for more than sixty years.

The study of Nature, as I have said, should be indirect, in order to perfect recreation. Some cheerful sport, to absorb one's direct attention, is the best aid to the end in view, and to my mind the best sport is that which necessarily takes one into the woods and along the streams, where wild flowers blow and wild birds sing, and where the flavor of sap and the fragrance of gums and resins are in the breezes. If I were a poet I think I should be one of that class described as

> " Poets, a race long unconfined and free,
> Still fond and proud of savage liberty."

I could not be the one of the garret and the crust; better a hollow tree and locusts and wild honey. The redeeming feature of Walt Whitman's deservedly tabooed, and yet deservedly admired, *Leaves of Grass*, is the sweet, ever-recurring wood-note, the sincere voice of Nature, half strangled as it is in incoherent sounds—a feature that affects one like the notes of a wood-thrush heard in the depths of a dismal, swampy hollow. Too much time spent in the streets and crowds of the cities— too much knowledge of the brutal side of life —has given us a Whitman, a Baudelaire, and a Zola. Too much knowledge of Nature gave us a Thoreau. It is a curious fact that, so

soon as a people have grown beyond the study
and the love of out-door nature, their literature
begins to be what French literature now is—
a literature without any true poetry.   Daudet,
for instance, is, a poet, but he cannot make
poetry.   His novels are spiced with intrigues
and immoralities, instead of with the flavor of
out-door life.   Zola sees nothing but the
tragedies of the gutter and the brothel.   He
never dreams of green fields and melodious
woods ; he finds nothing worthy of his art in
rural scenes or in honest, earnest life.   He
never goes into solitude with Nature.   The literature of England, from Chaucer down to
Dickens and William Black, is full of the fragrance, so to speak, of out-door life, and it will
be so as long as the English man and the
English woman remain true to their love of
all kinds of open-air pastimes.   The deer, the
pheasant, the blackcock, the trout, and the
fox, have done much to fence the poetry and
fiction or our mother-country against the
French tendencies and influences.

But American literature is beginning to
feel, in a certain way, the effect of much love
of Parisian manners.   Henry James, Jr., who
just now leads our novelists, is much more
French than American or English in his literary methods.   His theory is, that the aim of
the novelist is to represent life ; but he nowhere recognizes " out-doors " or out-of-doors
things as a part of life.   Life to him means
fashionable, social life—nothing more.   The
life of which Hawthorne wrote is *passé* to him.
From his stand-point he is right.   If realism,
as the critics now define it, is a genuine revo-
lution in literature, it may be a long while be-

fore any otherfiction than Mr. James's very pleasant sort will be in demand. He is master of his method, and has made the most of his theory. But, without finding fault with Mr. James's charming novels, it may be asked if they would not be better were it possible for the author to inject into them something of William Black's knowledge of out-door things, and to give them the color and atmosphere de. manded by the places where their scenes are laid. Social atmosphere he does give to perfection; but of the air his people breathe he knows nothing. He never sets his story in a landscape; its *entourage* is always an artificial one; he frames it, like an artist, with a frame exactly suited to its tone; but it would look as well in one place as another. In reading his stories we are thoroughly charmed, and would not know where to change a word; but we know all along that we are reading a story. He does not take us away from the spot where we are reading; but he chains us to our chair with the spell of his "representations of life" until the end is reached.

Now, a little different treatment would change all this. The color and the atmosphere of the place should be added, as with the brush of the painter, so that we would find ourselves on the spot, feel the air, smell the perfumes, see the varied features of the region round about, as well as talk with the people and share their life. Let it be understood that I do not criticise Mr. James. He is a prince of novelists. I merely attempt to show that he might add to his charming stories the freshness of the breezes, the bird-songs, and the flowers,

8

without abating in the least his placid realism
or endangering his reputation for merciless
analysis.

But even so delicately refined a novelist as
Mr. James loses less by the lack of a knowledge of out-door things than does the least of
minor poets. The singer must not, cannot,
rely upon any other reserve than Nature,
from which to draw the freshness and racy
flavor that every true poem must have. Still
it must be remembered that mere descriptive
writing, no matter how true to Nature, is not
what gives that " smack of Helicon " of which
Mr. Lowell speaks. The true critical test is
one that will discover any trace of the simplicity, the artlessness, and the self-sufficiency of
Nature. Whatever is truly fresh and original
in literature will be found to contain something
not acquired from books, nor from observation
of society, nor yet from introspection; this
comes, one might say, from the soil and the
air by a growth like that of the flowers. I believe it is due, in nearly every case, to out-
door recreation. It is felt on almost every
page of Emerson, Tennyson, and William
Black, and it is just as charming in a story
like *A Princess of Thule*, as it is in *In Memo-
riam* or in *Wood Notes*. John Burroughs has
shown what a delightful study Nature may
be to him who plays with her for the mere
sake of the play. He has given us the extreme of what may be called wind-rustled
and dew-dashed literature. What a grand
novelist Henry James and John Burroughs
would make if they could be welded together!
Life would then be represented sympathet-

ically from centre to circumference—from the heart of an oak to the outermost garment of a "dude."

Mr. Hardy's novel, *But yet a Woman,* and Mr. Crawford's *Mr. Isaacs,* leaped at once into popular favor on account of the freshness that was in them. In both stories a knowledge of out-door life is blended with a keen insight into the most interesting mysteries of the human heart. Mr. Isaacs was not only a master polo-player and a crack shot ; he was also a philosopher and a lover of no common sort. In *But yet a Woman* the descriptive passages and the epigrammatic paragraphs serve as a fixitive for the story, setting it permanently, and giving it an air of its own. The physical atmosphere is as wholesome and sweet as the moral spirit is sane and pure. One would suspect that the story had been written in the open air, or, at least, in the country, with the library windows wide open. Indeed, sunshine and air are as antiseptic and deodorizing in literature as in the field of physical operations. ·Even Baudelaire occasionally, under the influence of a sea-breeze, wrote such a poem as *Parfum Exotique,* or *La Chevelure.* He had a charming knowledge of marine effects, and it seems to me that his verse

"Infinis bercements du loisir enbaumé"

is enough of itself to immortalize him. It is a whole poem. One sees the warm, creamy tropical water, feels the long, lazy swell, the infinite idle rocking, the balmy leisure, and takes in, as by a breath, the illusive charm of the ever-mysterious sea. Buchanan Read's

Drifting might be condensed into that one line—

"Infinis bercements du loisir embaumé."

In fact, the few poems worthy the name, written by Baudelaire, were made out of the sweet, warm shreds of his out-door life, while on a voyage in the far East. Even in France, this freshness of Nature is recognized and relished. In *Numa Roumestan* M. Daudet has, as one might say, wafted the odors of Provence through the streets of Paris. The critics felt the atmospheric change, and went to the windows to see the mistral flurrying along the boulevards. So, in America, when Bret Harte and Joaquin Miller sent their stories and poems over the mountains and deserts from our far Pacific coast, it was their freshness—their woodsy, dewy, out-door flavor that recommended them. A happy blending of the bucolic with the latest fashionable tendencies —a welding together of the pastoral and the ultra-urban, made a great success of *An Earnest Trifler*. It would be easy to multiply instances. The proofs are perfect that the influences of out-door life upon literature are of the subtlest and most interesting nature. Whilst every one must admit the paramount importance of human life in every form of literary composition, still the side-light of out-door nature is absolutely necessary to the historian, the poet, and the novelist, and he who neglects it fails in one of the prime requirements of the best art. As well might the painter draw a group of figures without color, atmosphere, or background, and expect to win the highest fame, as for the novelist or the poet to depend wholly upon human actions and conversations for his effects. The moral of all this need not be appended. Out-door life is the great recreator and regenerator. Nature is steeped in the elixir which has power to freshen and renew our highest facilities. If " the proper study of mankind is man," still it is safe to say that sound lungs, healthy blood, a good appetite, and a clear brain, are indispensable to such study, and are to be had only by those who breathe pure air, digest their food, and read the human heart by the light of the sun.

A FORTNIGHT IN A PALACE OF REEDS.

WHEN you reach the top of the bold hill known as Cedar Loaf, you may see the Coosawattee River winding away, in a direction diagonal to the length of the valley below, sparkling and rippling between its dense fringes of canebrake. There are broad rifts in the forests of pine, hickory, oak and tulip, through which shine the grassy glades or miniature prairies, peculiar to the North Georgia region. The old Indian Ford, from which the serpentine trail of the Cherokees used to wriggle away like a snake, is still visible, its steep approaches having somewhat the appearance of abandoned otter-slides. Nowhere in the world, I believe, can such beautiful foliage be found as that wherewith the forests of this wild region bedecks itself in April. The young hickory trees spread out marvellous leaves, more than a span in width, and the yellow tulip exaggerates both foliage and flowers. The dogwood and sour-gum, the red-oak, the maple and the chestnut, the cherry, the sasafras and the lovely sweet-gum all flourish in fullest luxury of life and color. Wild flowers, too, of almost endless varieties, leap into perfect blossom early in spring along every hill slope and in every valley, pocket, and ravine.

Not far from Indian Ford stood the Palace of Reeds, built by Nature's own hand, on a low bluff of the river's east bank. We found

it—Will and I—while rambling in the valley,
and, by virtue of the right of discovery, quietly
appropriated it for our indwelling during the
fair weather of the delightful Georgian spring.
Imagine two wild plum trees in full sweet-
scented bloom standing twenty-five feet apart,
with a thick-leaved muscadine vine flung over
them like a richly wrought mantle. The boles
of the trees are gray and mossy, fluted like
antique pillars. The ground is flecked with
rugs of dark Southern moss through which the
violets and spring beauties have found their
way. The keen odor of sassafras and the
delicate perfume of tulip honey comes along
the air. You stand on the threshold of this
natural palace, and looking through the
tender gloom of its arched hall you see the
cool river flowing and singing on. There are
bees in the air, wild bees whose home is
in some great hollow plane-tree not far away.
You hear the dreamful hum of tiny wings.
You see the plum flowers shake and let fall
their golden pollen dust, and the·reeds, the
tall gold-and-green reeds, rise all around the
palace forming its walls. The earth is warm,
the sky is pure and cloudless. Deep in the
brake a hermit-thrush is calling. A vireo beyond the river quavers mournfully.

The Palace of Reeds was handsomely furnished with a mossy log for sofa, two camp-
stools and a low canvas table. An easel stood
for most of the day in the clear light of the
west, opening just above the babbling water.
It is worth noting, because now it is a fragrant memory, that the drawing-board was
of red cedar. The box of moist water-colors,
the bird-sketches, the portfolio of pencil notes,

the half-dozen well worn volumes scattered
about give a strange air to this woodland
bower.　No farm or plantation is in sight.
If you can hear any sound of busy human life,
it is the singing of some merry negroes pro-
pelling a corn-boat down the river.　Usually
these boats passed us in the night.　They
were a kind of long, low keel craft with stern
paddle and oars.　Midway of the boat were
heaped the white sacks of corn.　The tall
dusky oarsmen swayed to and fro singing
meanwhile some outlandish but strangely fascinating song.

Here by the flaring light of burning pine-
knots we read Keats and Theocritus, Shelley
and Ovid in turn.　Our concurrent studies
were not plainly congruous, rather conflict-
ing, one might think, for we studied Greek,
practised archery, collected birds-eggs, made
water-color drawings of plants and birds,
read poetry, boated, swam, practised taxi-
dermy, fenced with reed foils, fished for bass,
and cooked admirable dinners !　A little way
off stood our cabin, or rather, our hut, into
which a sudden shower of rain now and then
drove us.　When the nights were clear we
hung our hammocks in the palace, and slept
suspended in the perfumed breeze.　Often I
awoke in the small hours and heard the raccoons growling and chattering in the brake.
At such times the swash of the river had a
strangely soothing effect, a lullaby of fairy
land.

Will had a nocturnal habit.　He would slip
forth, when the moon shone, long after I had
gone to sleep, and the twang of his bowstring
would startle me from quiet dreams as he let

go a shaft at an owl or a night heron. Reading over some of the notes I made at the time recalls the charmingly unique effect of certain sounds heard at waking moments in those outdoor resting-hours:

The leaping of bass, for instance, plash, plash, at unequal intervals of time and distance, breaking through the supreme quiet of midnight, comes to one's ears with a liquid, bubbling accompaniment, not at all like anything else in the world. The mocking bird (*Mimus polyglottus*) often starts from sleep in the scented foliage of the sweet-gum to sing a tender medley to the rising moon. At such time his voice reflects all the richness and shadowy dreamfulness of night. It blends into one's sense of rest and becomes an element of enjoyment after one has fallen again into slumber.

Frogs are night's buffoons. " Croak, croak, croak," you hear one muttering, and with your eyes yet unopened and the silence and stillness of sleep scarcely gone from you, you wonder where he is sitting. On what green tussock, with his big eyes jetting out and his angular legs akimbo, does he squat? Suddenly " Chug ! " You know how he leaped up, spread out his limbs, turned down his head and struck into the water like a shot. You chuckle grimly to yourself, turn over in your hammock, and all is forgotten.

Then the screech-owl begins to whine in its tremulous, querulous falsetto, snapping its beak occasionally as if to remind the mice and small birds of its murderous desires. The big horned-owl laughs and hoots far away in gloomy glens. The leaves rustle, the river pours on, and the wind sinks and swells like the breath of a mighty sleeper.

Perfumes, too, affect one strangely, on waking, in the depth of night. There is a certain decayed wood in the Southern forests which at times gives forth a delicate, far-reaching aroma. This, together with the occasional wafts of sweet-gum odor and the peculiarly sharp smell of pine resin, steals through the woodland ways and touches the sleeper's senses until he slowly awakes. Drowsily he lies, ·with his eyes lightly closed, noting the tender shades of sweetness as they come and go. But the falling of a slight shower of rain, one of those short, light, even down-comings of large drops, which is not strong enough to break through the leaf-canopy overheard, moves the out-door slumberer to most exquisite enjoyment. He opens his eyes and all his senses at once. The air has sweet moisture in it, the darkness is deep. Above, around, far and near, a tumult is in the leaves. The shower is scarcely more than momentary in its duration, but it is infinitely suggestive. There are millions of voices calling from far and near. Vast organ swells, tender æolian strains, the thrumming of harp-strings and the exquisite quaverings of the violin. Multitudes clapping hands and crying from afar in applause. Then as the cloud passes on, the throbbing sounds trail after it, and at length it all dies out beyond the hills.

So our nights were " filled with music " in the Palace of Reeds.

Our days were the scenes of greater because more active pleasures. We had a pirogue dug out of a tulip log which we propelled on the river in our shooting, sketching and fishing excursions. We endeavored to make pencil studies of all the wild-birds in their natural attitudes, drawing them in water-colors afterwards from specimens held captive. These models we took in springes, traps, and snares of various sorts, the horse-hair slip-noose being the best for many birds. When the mulberries are ripening you may capture woodpeckers readily by erecting a smooth, slender pole projecting somewhat above the tree-top and having horse-hair slip-nooses, thickly set along its sides, for entangling their feet. The same capillary arrangement on the branches of trees especially haunted by any other bird will prove a pretty certain means of ensnaring it. We took great pains not to hurt our captive models and freed them as soon as possible.

Sketching a wild bird in the freedom of the woods and brakes is the utmost shorthand known to the artist. It must be done with all the dash and hurry of phonographic reporting. Five seconds cover a very long stop in a bird's movements, and some of them are never still for even that short period of time. I have followed one bird, a species of warbler (*Sylvia vermivora*,) for a full hour before I could get a passable outline sketch. In and out among the leaves, over and under and round and round, it went flitting, peering, prying, a very embodiment of restlessness. Such a chase has in it a smack of excitement, and after it is all over a leisurely survey of your sketch-book, leaf by leaf, will be both amusing and instructive. There is something of inspiration often found lurking in lines dashed down upon the paper in this hurried, almost frantic way. You have also sometimes made comic pictures when you least intended such things ! Here is a bird's bill and a quick firm curve for the back of its head ; the rest of the sketch flew away with the original. On the next page stands a fly-catcher, on one leg, minus a wing and having only the hint of a tail ; but you have preserved the characteristic attitude, and the sketch is valuable. You can work it up at your leisure. Here is a pine-woodpecker, a pretty fair outline, but there is no sign of an eye in the bird's head and its feet grasp thin air. All these notes, however hurried and uncertain, are reminders of what your eyes have seen, bringing up at once vivid pictures of the gay wild things which have flitted before you.

Sometimes a bird will be exceedingly accommodating. I recall now how one day I crept, under cover of a tuft of wild sedge grass, to within thirty feet of a log-cock (*Hylotamus pileatus*), and worked out a most satisfactory study, while it was quietly eating winged ants, as they poured from a hole it had pecked in a rotten stump.

. The yellow-billed cuckoo is a very difficult bird to sketch, so shy and sly and so restless. You will hear his queer, throbbing note in some lone place, and you will slip along hoping to see him. When you have nearly reached the spot, lo, he has eluded you, and his mournful voice caws out from deeper shades further off among the tangled trees. The wood-thrush and hermit-thrush are equally evasive. By the way, Wilson claims that the hermit-thrush is mute. I am sure this is an error. One day while I lay in a canebrake watching a green-heron's nest, a low sweet "*turlilee*," much like the wood-thrush's warble or thrill, called my eyes to a bird not ten feet away from me. I was well hidden and motionless, so that I was not discovered until after I had thoroughly identified the hermit. It repeated the low, musical trill several times, and when at length I frightened it by some movement, it flew away uttering a keen squeak or chirp.

Having digressed thus far it is pardonable to go a step further and declare that the blue-jay sings. I have heard it sing a low, tender wheedling song which seems never to have attracted the notice of naturalists. A wood-duck had her nest in the hollow of a plane-tree just across the little river from the palace. I watched her go out and in. She made her wings silent, so as not to attract notice, going through the air with as little noise as an owl. Her mate, a beautifully painted fellow, lingered about the brakes in the vicinity, occasionally uttering a sly quack. How the young when hatched were conveyed safely to the ground we failed to discover. One morning they were in the river swimming beside their mother as if they had always been there, doddling their heads and arching their necks just like old ducks.

There was an island a mile up the river whither we often went, to fish off shore for bass, and to sketch kildee-plover and sand-pipers. On one end of the island grew a patch of cane and rush-grass into which we tracked a fawn ; but the shy creature hid so successfully that we could not find it. A wild turkey had its nest in the edge of this jungle early in the spring. It was also the nesting-place of a pair of cardinal-grosbeaks, whose well-built home I discovered fitted neatly between three strong reeds. Soon in the morning the male would alight on the highest point above the nest and whistle bravely, his plumage shining like dull red fire.

There is no craft like a dug-out, that genuine Indian pirogue, for perfect gentleness and sweetness of motion. You sit on a seat hewn in the stern and ply a short, rather broad paddle. The long, slender boat is all before you, the prow well up, like a pug nose. The round, smooth bottom slips along almost on top of the water, as if running over ice. In such a pirogue we would paddle around the island and troll for bass, often catching wonderfully game fellows of over four pounds in weight. This silent gliding of the dug-out makes it *par excellence* the angler's craft. There is no rattling of rowlock and thole-pin, no oar-dip. Your paddle goes in silently, it comes out with not even the slightest ripple-break. The bass and bream are utterly unaware of your movements.

Speaking of bream, as the Southerners call the blue-perch, it is a royal fish. You find it in the eddies and swirls of those Georgian brooks and rivers, a voracious feeder, taking the worm with all the vigor of a trout. You use a rather heavy reed for a rod, rigged with a small reel. The larvæ of wasps and angleworms are the most killing baits. A bream weighing ten ounces will give you a lively run, testing your skill equal to a speckled trout of a like size. It comes out of the water shining with royal purple and yellowish waves of color.

In shape it is shorter and broader, but resembles somewhat the rock-bass.

We sketched our fish while alive, and I find, among many other curious reminders of the palace, a pencil drawing of the great Southern gar, a fish with a bill much like a snipe's. This specimen we did not catch, but bought it of an old negro, who, every Saturday, rain or shine, visited our camp, coming from a plantation quarter some miles up the river. He was a piper, a sort of African Pan, who blew lively pieces of barbaric tunes out of reed joints arranged in triangular form. He came to sell us eggs of the guinea fowl, which I suspect he stole, albeit they made very fine omelets. He taught us a new and ingenious method of snaring hares and birds. Our water-color sketches were wonderful to his eyes, and he babbled about them in a supremely droll way.

To dwellers in the Northern and Middle States, it may seem strange, this out-door life, but it must be remembered that the hills and valleys of Cherokee Georgia, are dry and warm from April to September, dews are light, the air pure, and, for weeks together, the sky is cloudless day and night. I recall a perfect February, it must have been in 1859. Will and I, then mere boys, staid out during the entire month and *not a drop of rain fell.* Every day was warm and clear, the nights were cool and pleasant. No clouds, scarcely any wind —a month of rare dreamy weather, not unlike northern Indian summer.

Many a night in July and August I have slept in the open air under a tree, preferring it to a cot or bed indoors. A hammock and a heavy blanket, for the nights are chilly even in midsummer, with mere shelter from dew if any fall, are all one needs for healthful rest.

Our bower among the reeds caught that gentle current of air which nearly always flows with the way of a river, and we were rarely disturbed by gnats or mosquitoes. There were no dangerous wild beasts, very few poisonous snakes, and, of course, nothing else to make us fearful.

But we were not idle dreamers. We had in view a definite object, toward which all our studies and labors pointed. Alas, the cataclysmal years which soon came swept all away! The best that can be gathered from fragmentary remnants and vivid recollections is a sort of dreamy pleasure in somewhat living over again those days and nights of tranquil greenwood life. A little of science and a great deal of nature we found out. We learned the ways of the fish, the birds, the bees, the winds, the clouds, the flowers. We translated the meaning of stream-songs and leaf-murmurs. In the Palace of Reeds we knew utter freedom based on older law than *magna charta* or any declaration of rights. When one is a supple boy in the wildwood, healthy, happy, strong, with a long bow in his hands and old romance all through him, he is free as the winds and birds. Add to this a strong purpose, an aim far ahead, and what would you have more?

Our indoor days, if those spent in the Palace may be so called, would have appeared, to a world-wise onlooker, somewhat tame; but to a poet they would have revealed the labors of sincere, earnest souls, feeling their way through youth's morning-mist to the clear light.

I remember one hot May day, too sultry for

any great physical exertion, we spent in the
most delightful way. Will was busy with Theocritus, and kept up a running comment on the
oral translation to which he was treating me,
while I, with leisurely care, was making a drawing in water-colors of a fine butcher-bird I had
captured the day before. The wind came in
desultory throbs through our mossy hall, fetch-
ing up from the river a touch of dampness and
the smell of water weeds. All the bird-voices
were hushed, or, if heard at all, they wasted
themselves in scattering squeaks and lazy
dreamful flutings. Shut away from the sun,
we were made aware of his extreme heat indi-
rectly by the softened reflection from the water
and by that dusky dryness always observable
on the reed leaves and the blades of aquatic
grass when a spring day burns like midsummer.
We could hear the chattering cry of the kingfisher and an occasional plash, as the industrious bird plunged into the river after his prey.
Diagonally across the stream, near the other
bank, a small tree growing at the water's edge,
had caught a scraggily drift of logs and boughs,
round which a brown scum, with huge pyra-
mids of white foam, was clinging. Some green
herons stood on projecting sticks, stretching
their puffy necks, or silently sulking, with
their sharp beaks elevated and their throats
knotted into balls upon their breasts. Among
some stones in a shallow place, a bright spot-
ted water-snake lay in the ripple, holding up
his angular head and darting his malign tongue
in sheer wantonness of spirit.

Those idyls, as Will read them, fell from
his lips to immediately blend with the warm
lull, the glowing dream of Nature. Those flow-

9

ers of song joined well witn the flower-de-luce
and the wild geranium. Their racy fragrance
was of kin to the leaf-smell and resin odor.
Will's voice seemed, in some mysterious way,
to become the expression of the mood of Nature. A dream came upon me. I leaned
against the wall of reeds and felt the coolness
of their sappy stalks steal all through my frame.
My sketch faded from my sight and I but
vaguely noted the restless movements of my
captive shrike.

There are times when hearing a true lyric
read aloud is the quintessence of all rapture-
ful music. It is the expression of everything
ariose and thrillingly sweet which has ever
been played or written or sung, from Terpan-
der to Remenyi, from Anacreon to Aldrich.
I said something of this sort to Will in reply to
a kindred suggestion from him touching the
idyls. He arose and strung his bow, then,
holding his ear close to the cord, he twanged
it softly and replied : " You hear that low note.
Well, how many ages ago did man first hear
it? The piano, the violin, the lyre, every
stringed instrument is a growth from the long-
bow. So some poet away off in yesterdays let
fall the first perfect seed of song, and its kind
will go on increasing in vigor and multiplying
in number forever."

Somewhere, in the depths of the brake, a
cat-bird began to trill and warble, and a big
bass leaped above the water of the river, beside
a half submerged log. The sun crept on and
rolled down the west. As the shadows length-
ened the heat withdrew, giving place to re-
freshing coolness. We watched the little
flurries of wind rimple the river's face. Great

turtles came up out of the water and crawled
along on a sandy place. Two doves circled in
the air, sailing like sparrow-hawks, getting
lower and lower, until they lit upon a stone
in the shallows below us and drank thirstily.
We heard the woodpeckers pounding in the
woods behind the hill, the nuthatches crying
" ank, ank," in the great tulip tree hard by,
and high overhead, in the yellow glory of sunlight, a hen-hawk screaming. Odors arose
and passed down the waxing wind. The cane
leaves tipped each other lightly, and a whisper-
ing of many voices arose from the rushes and
flags. So twilight thickened into night. The
stars crept out and the great horned owl and
the night-hawk crept out, too, with some solemn
bats and giant moths, that whirled and darted
above the reeds.

Such a fortnight in the woods as I have been
lightly sketching, will bring to him who rightly
uses it a rich return for whatever sacrifice it
compels. It is to Nature one must go for
ideas. Her lessons are rich with original
germs for the philosopher, the poet, the artist
or the romancer to vitalize his works withal.
No genuine bit of originality can be found, in
poem, picture or tale, which has not been
drawn from the secret depositories of Nature.
The woods and streams, the hills and winds
are but the indices to volumes, one leaf of
which would exhaust the literature of ages.
All eloquence, poetry, and painting can be
better understood when one is as free as the
winds and as happy as a brook. To know
what is supreme enjoyment, go into the woods
and, lying beside a rivulet in fair June weather,
read Theocritus till the bubbling stream and

the rhythmic idyls flow together in your mind a perfect harmony of naturalness. Or, if you are an artist, set up your easel by the brook, or, with sketch-book in hand, follow the vireo and wood-thrush from spot to spot until you have noted something new, if it be but a new attitude of the shy, shadowy things. Lie on the cool earth and watch the wind wave the trees and see the sunlight flit and flash through their high tops like rare thoughts through a poet's mind. Leap up and shout and sing. Take off your hat and toss your hair in the breeze. Plunge into the river and dive and swim. Go sleep in a hammock in the Palace of Reeds!

CUCKOO NOTES.

TAKEN at the right season, the mountainous region of northern Georgia will furnish a practically unworked field to the naturalist and pedestrian tourist, whilst to the artist it must become, sooner or later, a source of rich treasure. No other part of our country offers so pleasing a variety of landscape features, from the quiet repose of level river-fed valleys to the grandeur of rocky peaks thrown up against the bluest sky in the world.

This region is the Spring haunt of a large number of our American birds, as it affords the best possible nesting- and feeding-places for them, especially those whose habits are insectivorous and arboreal; besides, it is in the direct line of migration from Florida and other southern winter resorts to the great northern summer habitat of those happy feathered aristocrats who can afford to oscillate with the sun. The peculiarities of soil, the suddenness with which Spring comes on, and the protection to tender germs afforded by the curiously mountain-locked " pockets " and valleys, cause all sorts of forest and field vegetation to leap into vivid, lusty life early in April.

There is no word in our language so expressive of the sudden appearance of leaf and flower all over those brown hills and slate-gray valleys, as *gush*. The rains practically end with March, and the sun ushers in the succeeding month with a fervor that would be uncomfortable but for the ever-fresh breezes ; the light vegetable mould of the thin forests warms at once, and within a few days everything is green with leaves and gay with flowers.   Even the oak-trees have scarcely time to show their tassels before their leaves have broadened to dimensions wholly beyond comparison with those of oak foliage in any lower or higher latitude.   An almost dazzling vividness flashes, so to speak, from valley to hill-top, indicative of an exceptional local climatic impulse.   Everything grows with a *riant* haste, as if aware that this ecstatic Spring vigor would soon exhaust itself (as it nearly always does) and leave the region to a long, dreamy Summer drouth.

The migratory birds drop into this favored district, just in time to get the full benefit of its luxuriance, and are met by a clamorous and querulous army of residents, whose domain is too large to be successfully defended against invaders.   The wild orchards of plum and haw that border the glades, the thickets of young pines, the hickory groves and the dusky forests of post-oak and black gum are at once flooded with song.   The semi-marsh lands where the liquidamber * flourishes, and the river "bottoms" where the tulip-tree and the ash and elm grow to giant size, are the haunts of the pileated woodpecker, the hermit-thrush,

* The sweet gum (*Liquidamber styraciflua*) is a beautiful tree growing to perfection in the Southern States, along the banks of small streams in wet land. The gum or resinous balsam obtained by scarifying the bole is of a clear amber color, is pleasing to the taste, and gives forth a peculiarly agreeable odor. The tree bears a flat oval berry of a dark blue color much sought after by the golden-winged woodpecker.

and many another of the shyest and rarest of our birds.

Nearly all the rivers and rivulets of North Georgia are bordered with canebrakes and overhanging trees, darkly cumbered and bowed with the wildest masses of muscadine vines. The canoe-voyager passing down the Oostanaula, the Connasauga, the Coosawattee or the Salliquoy—streams as free and unconventional as the savages who gave them their musical names—will have exceptional opportunities for studying nature at first hand.

It was down these rivers that the rich planters, whose isolated plantations were scattered at wide intervals along the " bottoms," used to despatch their corn and wheat, their oats and cotton, in keel-boats manned by the happiest slaves who ever sighed for freedom. Many a moon-lit night I have lain on my bed of cedar boughs on a high, breezy bluff of the Coosawattee and heard those merry-hearted boatmen go by with the current, playing the banjo and fluting on the genuine Pan-pipe of graded reed-joints.* Recalling the music, at this distance, it seems to me the most barbaric and withal the most fascinating imaginable. Usually, no matter how bright the night, they had a fire of pine-knots flaring at the boat's prow, near which, on the rude floor of the forecastle, they

* This pipe is, in fact, identical with the *Syrinx* or Pan-pipe of the ancients. I have seen and examined many of them, formed of from five to seven reed-joints, of graduated sizes, bound together in a row. The music is made by blowing the breath into the open ends of the reeds. There were some reed-blowers among the slaves of North Georgia who executed certain characteristic negro melodies with surprising effect.

danced their vigorous hoe-downs, jigs and jubah-shuffles.

The hill country is, for the most part, very thinly settled, and many plantations once fertile and prosperous now lie waste, all overgrown with dew-berry vines and persimmon thickets. Everywhere, however, the birds find rich picking in the season of young leaves and larvæ, and all those perfumed and flowery groves are charming nesting-places.

Rummaging among my ornithological notes, I find enough material touching the habits and haunts of our American cuckoos to make a liberal volume. Most of the memoranda refer to North Georgia, and, in fact, the yellow-billed cuckoo (*Coccygus americanus*) especially, is more numerous there than anywhere else that I know of. The habits of this bird as well as those of the three or four other species found in North America, are extremely interesting, disconnected from any mere scientific view, and the places these birds inhabit, and the season during which they may be studied, make the pursuit of knowledge touching them a most delightful affair indeed.

The old nursery rhyme:

> " One flew east, one flew west,
> One flew to the Cuckoo's nest,"

should have read:

> " One flew *south* to the cuckoo's nest,"

in order to conform to American facts; for it is below the Cumberland range of mountains that one may find the paradise of cuckoos. Of course even the yellow-bill comes far North and nests in our apple orchards, forewarning

us of rain, as many good people think, by uttering its notably strange cry, once heard
never forgotten; but yet it is on the northern
margin of the sub-tropic, among the dry, warm
hills of Georgia, Alabama, and the Carolinas,
that *Coccygus* most loves to dwell.

A cuckoo's nest is a very simple affair—at
first glance, a mere amorphous jumble of
twigs, catkins and leaf-ribs, apparently tossed
at hap-hazard on a low bough; but it will bear
close study, for its architecture is characteristic of the bird's strange genius. How does
such a loose pile of sticks maintain its place
during a heavy wind? Careful examination
discloses a system of deftest weaving instead
of a careless or chance arrangement. The
work of a genius may appear rough and dis-
jointed when in fact the subtlest art has made
it look so for the deepest purpose. We may
never determine how near is the relation between the rarest human intelligence and the
instinct of animals, but I have not yet seen
the man who could build a cuckoo's nest!

From the Ohio valley down into Florida I
have tracked the cuckoo through all his seasons and haunts; but, as I have already said,
it was in the hill-country of North Georgia that
I made the most of my notes. Thither, therefore, let us go in the first days of April and be
on the ground when the strange, sly, shadow-
like bird comes up from the farther South.
He usually comes, with the wind in his favor,
drifting down into the fragrant groves on that
half-enervating, half-inspiring dream-breath
which the Spring puffs over the hills from the
gulf. The first notice given of his advent is
that pounding note, dolefully sounded in the

dusky depths of the woods, hearing which the old plantation negroes used to sing their watermelon rhymes :

" Plant yo' milions w'en de rain-crow holler,
 Ef yo' doan dey wont be wo'f er quar' dollar !
  Ki fo' de rain,
  Ki fo' de crow,
 Ye orter see how de wa'r milion grow ! "

It is not so remarkable, after all, that the cuckoo is called Rain-crow throughout the entire area of its habitat, for he seems always able to conjure up a shower within a day or two of his first appearance in the spring. I suspect that he holds his solemn voice until the rain is at hand, so as to make a fine artistic unity out of it and the depressing gloom of a rising storm-cloud.

The haw-groves that usually fringe the margin of the mountain glades are the Yellow-bill's favorite resorts when it first reaches the hill-country from the south. Here it meets the blue-jay, the brown-thrush and the cardinal-grosbeak, permanent residents and implacable claimants of all the fruits and insects of these favored spots.

A glade is a peculiarly Southern woodland feature, not found in perfection north of Tennessee, a miniature prairie, surrounded by scrubby trees and groves or thickets of plum and haw-bushes, and covered, as a rule, with wild wire-grass and tufts of sedge. Every one who has spent much time in the wildwoods has noted how few are the small birds inhabiting forests of tall thickly-growing timber ; but these glades, set in the midst of immense tracts of pine and oak woods, are oases of

bird-life, as one might say, where in the singing season the air is shaken with a sweet tumult of voices. Here the persevering egg-
collector is sure to find the delicately-tinted
treasures with which he delights to decorate
his cabinet. The butcher-bird, the grosbeak,
the cat-bird, the wood-thrush, the brown-
thrush, the robin, the blue-jay, the mocking-
bird and the cuckoos all like to build their
nests in the thorny arms of the haw and plum-
trees. All these birds are, in a degree, bit-
ter foes of each other, allowing no opportunity of venting a little spite to go by unim-
proved, but they rarely go to the length of
committing any irreparable wrong. True, the
blue-jay now and then robs a nest and the
shrike may impale a smaller bird on a thorn,
but these acts are the rare exceptions in the
mating and nesting time.

The cuckoo, however, must be closely
watched by all the rest or it will slip its egg
into a stranger's nest. Our American bird is
very sly in performing this parasitic trick, so
common to the European species, and is
guilty of a sin in connection therewith which
adds greatly to the ugliness of the main crime.
I am led to believe, on the strongest circum-
stantial evidence, that the yellow-bill species,
at least, not only carries its egg to the nest of
another bird, but that it also invariably takes
away from the nest one of the eggs rightfully
there. This habit is a very curious and interesting one. Our cuckoo always builds a
nest of its own and rears its brood with ex-
emplary care. The eggs it scatters on occa-
sion here and there in strange nests are probably the result of over-fecundity, for at best

it appears to be erratic in its laying, the eggs in its own nest varying greatly in number and in development stage.

I have collected and arranged all the obtainable facts on this subject, and my conclusions in short are: That the cuckoo of North America, more especially the Yellow-bill, may be either slowly losing or slowly gaining the egg-depositing or parasitic habit of the Old-World species; that it is exceedingly eccentric, in connection with this habit, acting from the impulse of accidental necessity on account of an irregular fecundity. Its nest-building habit will not admit of its rearing a large brood of young; its eggs must, therefore, be divided among the nests of its neighbors: that is, whenever the over-supply comes on. The bird itself, as regards the two species (black-billed and yellow-billed) with which I am well acquainted, is very strangely sly, furtive, and erratic in all its actions, affecting a close observer with the impression that it is all the time laboring under some restrictions or limitations not common to birds in general. Its movements are graceful, but there is in them something that suggests unsubstantiality—the lightness that comes of an ill-balanced nature.

Its form is elongated and so accentuated by its slender, curving bill and disproportionally developed tail, that it appears almost serpent-like at times, as it creeps with a noiseless gliding motion through the foliage. There is never any evidence of happiness in its actions or in the sound of its voice. On the contrary, the cuckoo appears to be the embodiment of aimlessness, restlessness, and unmeaning discontent. Its solemn, almost gloomy, hazel eyes, and the peculiar way it has of glaring half-stupidly at one when one approaches it, adds much to this unbalanced effect. In flying from one tree to another it does not cut straight away through the air, but dives downward, nearly to the ground, sometimes, and then whirls along in a zig-zag, erratic line, rising again at a sharp angle before alighting. While in the air there is a sparkle of white in its over-long tail, and a sheen of greenish silver-gray along its neck and back, while on its wings trembles the glint of burnished copper blended with reddish cinnamon tints.

While in repose it may be described as follows: Bill black above, yellow below, long, broad at base, gently curved; feet lead-colored; back, darkish olive-gray; under parts, white; wings shot with vivid cinnamon, especially on inner webs of quills; tail bearing on central feathers a continuation of the color of the back; outer tail-feathers tipped and edged with clear, pure white. Total length, 11.50 inches; alar extent, 16.00 inches.

Its nest when built in an orchard differs in construction somewhat from its wildwood architecture; but it may be easily identified by the open, sketchy effect of its outlines, its flatness and shallowness and the presence in its texture of the tassels and spikes of amentaceous trees carelessly woven through the tangle of coarse twigs and fragments of leaves. The eggs, deposited irregularly in the oval, saucer-like cup, are of a very delicate greenish shade of color not easy to describe. I have found occasionally as many as seven in a nest, though four is the usual number.

Our cuckoo is not an "egg-sucker," so far as my observation goes. The popular tradition giving him that villanous habit, has arisen, no doubt, from the fact that he has been seen with an egg in his mouth. I can think of no wildwood effect more likely to gain a lasting lodgment in one's memory than the appearance of this bird flying along with an egg between its mandibles, seeking some other bird's nest in which to safely lodge this surplus fruit of an erratic habit.

The Black-billed species (*C. erythrophthalmus*) is a little smaller than the Yellow-bill, and far less singularly interesting. It lacks the white sparkle in the tail and the bright reddish copper wing-glint, as well as the dash of yellow on the lower mandible ; otherwise it is much the same in appearance with *C. americanus.*

I once had a bush-tent built of fragrant pine and cedar boughs at the margin of a glade, not far from the bank of the Coosawattee, where I spent a fortnight in the systematic study of the yellow-billed cuckoo, the lesser shrike, the mocking-bird and the catbird. This period extended from about the 10th to the 25th of April. All around the glade grew honey-locust trees, haw-bushes, crab-apple and wild-plum thickets and dense tangles of blackberry vines. Everything was heavy with leaf and bloom ; fragrance loaded the air, and the birds all appeared in a great hurry to build. I could sit in my tent door during the dewy morning hour and watch the love-passages, the quarrels, the fights, the nesting troubles and triumphs of these gay things with not a waft from the busy human world to disturb my enjoyment.

A pair of yellow-billed cuckoos were building a nest, after their desultory, aimless fashion, in a scrubby tree over which a mass of the Southern green-briar vines had grown. The bough upon which the beginnings of the nest-skeleton appeared, was not more than forty feet distant from my door, so that, barring some slender intervening twigs, I had a clear view of all the building processes. One curious and noteworthy habit of the cuckoo was observed, of which I have never seen mention in any ornithological work. In carrying a limber twig or leaf-fragment, the bird gripped one end of it with its foot and the other with its bill ; a trick which enabled it to pass through the tangled vines and branches without much difficulty on account of its burden.

During my stay at this glade the nights were · rendered glorious by a strong moon and a clear atmosphere. Several times I heard, between midnight and dawn, the cry of the Yellow-bill uttered in a suppressed tone from the densest part of a thicket. It may have been a mocking-bird. I tried in vain to be sure, but I am inclined to think that the cuckoo itself uttered the calls. If it was a mocking-bird the weird reserve-force apparent in the expression and *timbre* of the imitative passage did infinite credit to the famous low-country songster's incomparable vocal powers.

It is strangely difficult to make out the exact location of a bird by its cry at night, especially in a wooded place. I tried to discover the roosting-place of my cuckoos ; but watch them closely as I could, they evaded me. They appeared not to care much for each other's company, save when in a loving mood, and I think they roosted without any reference to companionship. Early in the morning, however, the pair found each other out, and joined in the labor of nest-building or the pursuit of caterpillars and other leaf-eating insect forms, with a reasonable show of conjugal unity of purpose.

Their nest progressed very slowly and jerkily. Now and again, for two or three days together, nothing was done to it, then for two or three hours the work would be unceasing. They behaved themselves after the manner of awkward and not very apt tyros in the art. The male was even silly in some of his performances, time and again carrying away from the nest a stick (that had previously been worked into it with great labor and care), apparently in a fit of absent-mindedness.

This unaccountable listlessness or characteristic oddity of behavior is not confined to the genus now under consideration, but runs like a family taint through the whole catalogue of *cuculidæ*. The ground-cuckoo (*Geococcyx californianus*) is an embodiment of drollness and absurdity. The Ani (*Crotophaga ani*) is another very interesting kinsman of our bird ; but instead of scattering its eggs among the nests of other families it has the opposite habit, several females laying their eggs and together incubating them in the same nest !

The *Cuculus canorus* of Linnæus, which is the cuckow or cuckoo of England and Africa, has attracted more attention than any other bird in the world. Some very strange facts touching its history have been gathered. It would indeed fill quite a volume if one should give only a compendium of cuckoo literature, most of which refers to the European bird. Quite a discussion was precipitated into scientific circles when, some thirty years or more ago, a distinguished gentleman propounded the statement that a cuckoo invariably colored her egg to coincide with those in the nest chosen as the place of deposit. A cabinet of eggs, claimed to be those of the cuckoo and those of the birds in whose nests they had been found, was arranged for the purpose of demonstrating the apparent truth of the startling theory; but notwithstanding many curious facts, it could not be maintained.

It remains pretty well settled, however, that the eggs of *Cuculus canorus* may now and then vary in color, somewhat in accordance with the hereditary individuality of the particular bird, and that each female cuckoo may instinctively choose, as a rule, to deposit her egg in a nest with those of a bird laying eggs of nearly or quite the same color.

So eccentric and variable is the Yellow-bill in its habits, that it is not at all wonderful that much doubt has existed as to whether it is parasitic; but I am convinced that it does, irregularly, under stress of over-fecundity, slip an egg occasionally into the nest of another bird, and this habit and others characteristic of the genus, appear to be imperfectly formed as yet, or else they are being gradually abandoned.

This apparent tendency towards sloughing hereditary habits, or acquiring new ones, is noticeable in several of our American birds,

notably in some species of woodpecker; but
our cuckoos are the best instances for study.
A good binocular glass and a season or two
of patient observation will enable any intelligent person to detect a great deal of evidence
of this tendency in cuckoos.   The yellow-
billed species carries its vacillating nature
on its sleeve, as it were, and forces it upon
consideration.   The black-billed species is
scarcely less peculiar at most points ; if there
is a difference it is of degree only.   Even the
ground cuckoo (*Geococcyx californianus*), is
almost absurdly peculiar and *outré* in its
habits.   Dr. Coues says :   " They are singular
birds—cuckoos compounded of a chicken and
a magpie ! They prefer running on the ground
to flying, only using their wings as auxiliary
' outriggers ' while darting .along at almost
race-horse speed."   Dr. Coues notes in the
nest of this species the same slightness and
apparent awkwardness of construction so
marked in all cuckoo nests, " As if," he says,
" the birds were just learning how to build."

Our Yellow-bill may be taken as the strongest type of this strange family.   Haunting our
bloom-burdened and odorous Spring groves,
like some restless spirit of remorse, furtively,
dreamily, but ever with a look of suppressed
pain, it has affected the popular mind as if
with a superstition borne upon its own wings
from some undiscovered country.   Its voice
is considered ominous not only of rain and
storm, but of evil in all its mysterious and
undefined forms.   Of course this is an idle
popular delusion ; but it serves to point out
the exceedingly well-defined power resident

in any form of mystery, even if but the *quasi-*mystery of a cuckoo's ways!

Indeed, the bird, its habits, its individuality and eccentricity of nesting and of oviposition, and its half-mystified expression of the eye, its hesitating, skulking flight, and its evident lapses into absent-mindedness, may well serve to impress one's imagination, at least, with a suggestion of a transition state through which Cuckoo is passing to a lower or higher grade of character.

One day, as I was going down the Salliquoy, a small tributary of the Coosawattee River, I saw from my pirogue a cuckoo's nest on a low branch of a water-oak. The female was crouching on the insecure looking pile of sticks in utter terror, while a whole pack of blue-jays were screeching and fluttering in the foliage above it. I shall not soon forget the expression of that bird's great solemn eyes. Evidently the poor thing felt that a dreadful fate was impending over it. But the fact was that the blue-jays were worrying a little screech-owl that had ventured into the day-light, and which was now cowering in its stolid way on another branch of the tree near the nest.

Our Cuckoo, though not notably combative, will fight with great fury in defence of its young, and the males engage in fierce silent struggles for supremacy during the early part of the mating season.

The nesting area of the Yellow-bill extends from Florida to Michigan, and from the Atlantic coast to some line west of the Mississippi River, and I am inclined to regard the black-billed species as having nearly the same limits

of habitat. To what distance Canada is in-
vaded by either or both seems left in some
doubt.

Whilst the cuckoos of eastern North America are technically frugiverous, they are not,
so far as my observation serves me, strictly
fruit-eating within the general and popular
meaning of the term. I have never seen
either of the two common species taste any of
the small fruits, wild or tame. They probably
eat seeds at need, but their chief food is insects—the caterpillars, moths, butterfly-eggs
and various larvæ found on the leaves and
branches of trees.

The Cuckoo's habits may be studied to
advantage by any one who will take the trouble
to scan with care almost any apple-orchard in
Spring and be guided to the bird by that half-
solemn, half-comical cry uttered at intervals,
which may be phonetically rendered thus :
" *Kauwk, kauwk, kauwk kuk—kuk—kuk—kuk
—k—k—k—k, kauwk, kauwk, kauwk !* " In
uttering this singular call or cry, the bird begins slowly, the two or three leading notes
coming forth at nearly equal intervals, then
the succeeding ones are produced with rapidly
increasing quickness, until they run together
into a sort of rattling noise, succeeded by a
repetition of the opening cries. Loud, harsh,
peculiarly doleful, the voice of the rain-crow,
as the bird is vulgarly called, rings through
our woods and orchards, more especially in
cloudy weather, with an accent far from cheer-
ing or pleasing. Hence has arisen the unwar-
ranted ill-feeling existing in rural districts
against this very best bird-friend known to our
farmers and fruit-growers. The cuckoos

should be protected and their propagation encouraged, as they are the saviours of our forests, our orchards, and our hedges.

Looking over my cuckoo-notes, I find reminders in them of all the sweetest woodland solitudes between the Great Lakes and the Gulf. The bubbling of the cold trout-brooks of the Leelenaw blends with the lazy swash of the Pearl River and the Kissimmee.

But I must hasten to remark that, contrary to what one is led to expect, in all the low country of the South the cuckoos are scarce, even in mid-winter. In the region of Lake Okeechobee and on the outskirt of the Everglades close observation failed to certainly note even the species *C. seniculus* or mangrove-cuckoo. From the fact that the Yellow-bill is found on the Pacific coast and in parts of the Southern Rocky Mountains, it is probable that its winter resort may be chiefly in Mexico and Central America. In March I saw a few specimens haunting the oak groves on the high-lands between Tallahassee, Florida, and Thomasville, Georgia, and I was told that their nests were sometimes seen there.

So many cuckoo legends have gone afloat —each adding something uncanny or romantic to the popular opinion of our harmless bird —that I am tempted to close this paper with one current in the southern mountainous region, to the effect that the Yellow-bill cannot be killed by a rifle-shot if its breast be turned towards the shooter. I once attempted to demonstrate the fallacy of this claim for the benefit of a hard-headed old mountaineer and was unlucky enough to miss my bird !

" Ther' ! " he exclaimed, " what'd I tell ye !

ye mought es well shoot at a ghost, er a spirit."

The secret of the matter is that the cuckoo's breast is sheeny white and presents a very slender mark, which on account of its being just the color of the silver fore-sight of the common rifle, is very difficult to "draw a bead" upon; wherefore even the most expert marksman is apt to miss it.

SOME MINOR SONG-BIRDS.

OUR interest in wild song-birds must in-
crease apace with the narrowing of our wooded
areas, and in proportion to the constant lessen-
ing of our opportunities for ornithological
study at first hand. As our thrushes and
orioles and warblers one by one take flight,
we suddenly, in realizing our loss, feel in a
new way the sweetness of their voices. When
we were children, even if we lived in the heart
of the city, we often had glimpses of the
country with its great dense woods and its
green fields, its orchards, and its cottages
covered with morning-glory vines. In those
days the brown-thrush, the cat-bird, and the
cardinal grosbeak, sang in every thicket and
throughout every orchard. Now these charming little lyrists are gone from many a former
haunt ; indeed there are wide areas of country,
where they used to nest and sing, in which
they never will be seen in a wild state again.

Not long since I returned, after twenty
years' absence, to a neighborhood in which
my infancy was spent. I remembered a certain brook in a little field, a crooked lazy little
stream bordered with yellow willows and water
hazel, where the cat-bird loved to swing and
sing in shade and sun. It was with an indescribable regret that I found the willows and
hazel all gone and the brook, sunken under
ground, groping its way through tubular tiles.
Where wide woods of beech and sugar-trees

used to be, fields of wheat and corn lay green and smooth almost to the horizon's rim. What a loss the absent birds were felt to be! In fact, when, after much plaintive sauntering over the altered grounds, I chanced to hear a lonely purple finch twittering in a hedge of *bois d'arc*, I felt a thrill of delight which was like an electric message from my childhood's days. In the streets of the village which had shrunken, as if in some mysterious proportion to the widening of the surrounding plains of agricultural thrift, foraged a well-fed flock of detestable English sparrows. This, I thought, is advanced enlightenment—a covered ditch for a brook, a prim hedge in lieu of a wild plum thicket, an orchard displacing an odorous grove of wild crab-apple and these pests of sparrows usurping the homes of the cardinal-bird and the thrushes!

From almost any little country town, even in the West, one must now, as a rule, make a long flight into the most neglected nooks of the rural neighborhoods, before one can find the haunts of the more interesting songsters. The elect few of the feathered choir, like the choice spirits of the outer circle of young poets, are fond of utter, limitless freedom; they do not relish the fragrance, however sweet, of over-cultured gardens and bowers. True enough, the blue-bird warbles very contentedly on the best kept fence-row as he watches the ploughman turn up the tid-bits from the furrow; and it is an almost savage tenderness that quavers from his throat as he pounces upon the dislodged worm, his wings gleaming like some precious, doubly purified

gems fresh from the fabled fires of the empyrean.

Reading the above sentence over, I feel its
coarseness in the presence of a genuine blue-
bird-sheen and blue-bird-warble reaching me
as I write. How artificial and insincere are
the verbal rhapsodies of the most natural of
our poets when set in the searching light of
unconscious nature! Why do not the blue-
bird's notes, arranged always in the same
order and expressed always with precisely the
same tone, accent, and emphasis, become
stale? Why does not the bird's manner grow
perfunctory? Who ever did get weary of hearing over and over, from day to day, spring
after spring, those liquid bird-phrases that,
pitched to a strange minor, have been the
same since first an oscine throat was filled
with music? We must all, even the most un-
imaginative of us, acknowledge a little impulse
to gush and get rid of a fine fury of sentiment
about the time when a flash of green, a thrill
of warmth and balm, and a gush of bird-song
go across the fields and woods.

The man who can look into a bird's nest,
well-set with tender-hued eggs, without feeling
an inward smile, as if his soul were sweetly
pleased, has lost something that is the chief
ingredient of perfect sanity and simplicity.
What is usually meant by the word sentiment-
ality is an abomination; but our human nature, in a state of absolute health, is furnished
with a myriad little well-springs of generous
sympathy and sweet responsiveness, that
should not be allowed to go dry. If the fragrant, essential elements of a healthy soul
may be called sentiments, then let sentiment-

ality bubble like brooks in spring and gush like the thrush's song in nesting time.

Bird-hunting and bird-loving folk get the very best out of life in the way of sensuous pleasures not in the least voluptuous or over-stimulating. Just now, looking back over my notes, observations and recollections of out-door life, my long association with most of our minor song-birds appears something well worth having experienced. Much of what I remember is knowledge of a kind scarcely communicable by any literary or artistic means, or by any method of natural expression. Once I heard a blue-jay sing as sweetly as the mock-ing-bird when trilling in a tender minor key. I could hardly believe my own sight and hear-ing as the beautiful, tricksy creature sat before me with drooping crest and half-raised wings, swaying his body lightly up and down and uttering a low, almost bewildering flute medley, full of the cadences of dreams.

Still the blue-jay is not reckoned among the singing birds by those who are not close ob-servers. His common notes, though occasion-ally musical and sweet, are, as a rule, harsh and ill-tempered; a very imaginative person might conclude that the dolefully tender song I heard was the result of a fit of remorse, on the blue-jay's part, for myriads of sins com-mitted against the nests, the eggs, and the young of other and weaker birds. How often I have witnessed acts of the most brutal cruelty done by the jay in apparently the quietest mood imaginable!

I recall an instance now: A sparrow had a nest with young in a clump of lilac-bushes on a lawn in front of a room I was occupying.

One morning about sunrise, as I sat by a window, writing, I heard the mother-bird "chipping" dolefully, and I looked out just in time to see a blue-jay kill, by a deft turn of its powerful bill, the last remaining fledgling of the brood. The assassin then proceeded to tear up the tiny nest, after which he very perfunctorily flew away ! Here was a case of utter depravity—a piece of unmitigated outrage for which there could have been no motive aside from the impulse of a viciousness incomparable. I went to the spot, and found the young sparrows scattered on the ground, dead in the midst of the shreds of the nest. Each bird bore the livid pincer-like impression of the jay's beak. I cannot account for this well-known brutality of the jay ; it does not appear to be always present with it, for I have known it to live in perfect peace with other birds, nesting in the same orchard and even in the same tree. Its colors and its restless activity make the blue-jay one of the most valuable elements of almost every bit of thicket or hedge throughout our Middle and Southern States for nearly the whole year.

I am aware that many objections may be urged to putting so harsh a screecher in the catalogue of music-making birds; but it can and does occasionally sing most superbly. Moreover, upon being dissected, the blue-jay's throat shows a very high state of development, the muscular arrangement of the lower larynx bearing every sign of great flexibility and of delicate adjustment. It is a hardy bird, often met with in midwinter far north of the fortieth degree of latitude, apparently quite happy among the sleety and snowy branches of the

leafless trees. It is a good nest-builder, and
provides for its young with a great show of
affection and industry It customarily keeps
near the ground, but I have observed large
flocks high up in the air, migrating southward
in autumn.   .

Turning from a provokingly dual subject—
the paradoxical nature of our jay—one feels
relieved in speaking of the genial and melodious life of the brown-thrush. Next to the
mocking-bird the most famous singer of our
woods, this beautiful little fellow, with his
snuff-colored coat and dappled vest, is welcome
wherever he goes. My observations of his
habits extend over a wide area reaching from
Northern Indiana to Florida, and I have no
vicious trait of his character to record. In
the mountains of East Tennessee, and among
the hills of North Carolina and Georgia,
brown-thrushes are almost as common as are
blackbirds in the flat fields of Illinois. The
thickets that rim the glades, especially the wild
orchards of haw and crab-apple, plum and
honey-locust, are the favorite nesting-places
of this bird; but he chooses the topmost tuft
of the tallest tree for his perch while singing.
His song, full-toned, loud, clear, varied, is
often mistaken by casual listeners for that of
the mocking-bird, though really far inferior to
it in both volume and compass, and scarcely
to be compared with it in purity of resonance.
In the far South, where all birds are given to
greater latitude of habit than in the North, the
brown-thrush now and then sings in the night,
a low, dreamy, lulling song, warbled as if with
a sleepy ·throat. In this too he follows, but
does not equal, the mocking-bird. I have

habitually slept in a hammock while outing in the Southern woods, and no words can convey the singularly delicious sense of calm and quiet luxury which comes of hearing, far in the solemn night, the low, liquid, drowsy nocturne of one or both of these charming musicians !

The brown-thrush has not had his full meed of praise from our poets. As a conventional figure, the nightingale—a bird quite unknown to Americans—has retained its place on the palette of our word-painters, much to the hurt of our poetry. In fact, I fancy I can go through American poetry and point out every passage wherein an author has alluded to a bird that he has never seen. How can any one describe the fragrance of sweet-clover without having it in his nostrils at the moment of writing ? How can I write sincerely about the song of the brown-thrush or the cat-bird, if I have not the stimulus of that song in my brain ?

In the far-reaching tangles of wild grape-vines, found here and there in the beautiful little valleys of North Georgia, the brown-thrushes sing to the perfection of their powers from the early days of April until the first of June ; that is, they make the vine-masses their home, and do their melodious gushing on the very topmost boughs of the highest trees. This is not over-statement ; it is one of the most striking sights of the Southern woods to see a brown-thrush at about sunrise, sitting on the apex of the cone-shaped top of a giant pine-tree, whilst its song falls in a shower of fragmentary and ecstatic trills and quavers over all the surrounding woods. This performance often extends over the space of an hour or more, with but slight intermissions.

The nest of the brown-thrush is a straggling mass of twigs, roots, bark, leaves, and weed-stems, carelessly tumbled into a crotch near the ground, or on the flat projection of a fence-rail, sometimes even on the ground. Its eggs are delicately pretty, whitish or pale green, flecked thickly with brown, from four to six in number.

North of the Cumberland Range of Mountains, the brown-thrush is migratory; but in parts of Tennessee and North Georgia I have found it a permanent resident, especially in the brushy valleys. It is a hardier bird than the mocking-bird.

The cat-bird (what a name!) is one of the finest singers in the world—beautiful, too; but, for some mysterious reason, under a ban of disgrace and contempt throughout its wide habitat. You may know him by his dark slate-colored coat and gray vest, his black cap and chestnut-brown under tail-coverts, as well as by his peculiar cat-like mew when irritated. He is a lyrist of the dense thickets and brier tangles, the musical deity of our blackberry jungles and *bois d'arc* hedges. His song resembles that of the brown-thrush, but it is slenderer and keener, trickling through the leaves in a tenuous stream with ripples as light as air.

The nest of this species is well constructed, hung low, and its eggs are of a lovely deep greenish blue.

The cardinal-grosbeak is one of our American songsters, which, though much persecuted by fanciers and imprisoned in cages, is not justly appreciated. His brilliant red plumage and smart manners have been much better studied than his sweet and powerful vocalization. His notes are few, but the compass and volume of his voice, and the vivid force of expression he commands, are without rival. Not even the mocking-bird can equal him in his one circle of execution. He sings with true American energy, flinging out his notes as if from a clarion. His attitudes are those of unbounded self-confidence; he appears to claim the whole world as his own, as he stands bolt upright on a bough, his crest erect, his bold eyes flashing, and his voice leaping out with the impulse of a diminutive steam-whistle. He is a wary, shy, swift bird, but his color exposes him to the watchful collector, who is ever eager to take him. The cardinal's nest is well-built, usually set in a tangled place of a thicket. Its eggs are of a mottled reddish-brown color.

In the region of Tallulah Falls I met with an old man whose chief business was snaring red-birds (cardinals) for the sake of their skins, which he sold to a New York firm for use in millinery decorations. Most of his work was done in the mating season, when with a trained decoy-bird and a cage furnished with side-springes, he took great numbers. The method was to hang the cage, of open wire-work, with a live male bird in it, on a bough in the midst of a thicket. The springes at the sides of the cage were so arranged that no sooner did a visiting bird alight thereon than he was caught. The captive left alone calls loudly and is answered by a female who comes near. This excites the jealousy of her lord, who dashes at the cage and dies. The old man had four of these murderous contrivances, and was reaping a considerable profit from them. He understood his business perfectly, going about it with great energy, but evincing no enthusiasm or especial feeling of any kind.

In the thickly settled States of the West the orchards and hedges are, in spring-time, the abodes of many singing birds. The field-sparrow is chief among these, showing off his exquisite vocal gifts about the time that the young wheat is ankle high. His life is mostly spent on the ground where he runs through thick grass or cereal sward with a rapidity like that of the ousel in water. When the lyrical mood comes on he mounts to the top of a stump, a hedge or a fence, and pours forth a very sweet little carol, meantime elevating his head to the full extent of his neck, and puffing out his little throat after the manner of a toad.

The orioles and some of the warblers have cheerful voices, but can scarcely be called fine singers. They give a dash of freshness to our groves when they arrive early from the South, and, like our blue-bird, are always welcome.

Speaking of the blue-bird, he is uniquely American. He has no kin on the other continents. He appears to be a flake of the cerulean above, let fall, by a special dispensation, upon our favored country. Like some poets, he is always just about to sing, but never does more than begin his song. His fragments are divine, however, suggesting a reserve of something too sweet and fine for the common winds to bear. His is a rhythmical nature, and his flight is a poem in itself. As he goes trembling and wavering along through the air and sunshine, he adds to a May-day just the touch that makes it perfect. The blue-bird in its nest-habit offers for our study one of those curious contradictions now and then appearing in nature. Instead of building a graceful nest, swung airily amid the fragrant foliage, it dives into some gloomy, unsightly hole in a rotten stump or tree, and there, like the kingfisher in his subterraneous cavern, rears its brood. Querulous, saucy, bold, this beautiful little creature has endeared itself to every observer.

Our indigo-bird, bluer than the last-named singer, and almost as common, has attracted comparatively little attention. Its song is really fine, though delivered without expression, or any show of interest. One must approach very close to get the full sweetness of the frail, faltering strain which can be heard but a little distance. When it is caught in its completeness, however, the melody is so childish and tender that one forgives the inartistic manner of the delivery. The scientific name of this bird is *Passerina cyanea*, the specific part meaning dark-blue, and it may be identified easily by that color covering its head and shimmering with a greenish gleam over its back. Its nest is rather sketchy, built with little care, and set in a low bush, usually at a crotch. Its eggs are bluish white, sometimes slightly freckled.

With a word about Wilson's thrush I must close this paper. To my ear this bird's voice is purer and richer than that of the famous wood-thrush. Its shy habits, and the chary parsimony with which it doles out its vocal favors, have, no doubt, tended to prevent its becoming popular, even with good observers. There is a silvery ring in its higher notes and

11

a watery gurgle in its lower ones, that give to
its song, usually heard in low, heavily wooded,
dusky semi-swamps, a peculiar vibration altogether indescribable. Its nest is a curious
mixture of sticks, leaves, grasses, and rootlets,
usually set on or near the ground. Its eggs
are greenish blue. Of all the thrushes this
appears to me to be the shyest and wildest,
and while its voice lacks that flexibility and
compass possessed by those of the brown-
thrush and the cat-bird, it certainly has the
advantage at the point of *timbre* and of liquid-
ity. One can imagine nothing to compare
with some of its notes, unless it would be the
blending of the tones of a silver bell with the
bubbling of a brook over pebbles. Its song is
usually heard at a considerable distance, in
the twilight gloom of damp woods, and there
is a touching trace of melancholy in it that
makes it blend well with the environment.
Along the Wabash river, in the broad, wooded
" bottoms," I have heard it singing long after
sunset, and its voice is the first sound that
breaks the silence of the morning there.

One who has loved the woods and fields and
has spent much time in the pursuit of knowledge in the wild paths of nature, can look back
upon the days that are gone and see so many
bright visions—hear so many sweet sounds
and feel so many thrills through the nerves of
memory! One can scarcely be called senti-
mental if one gushes a little over one's sweet
experiences.

The next best thing to having cheerful and
healthful memories is the liberty of imparting
something of their effect to others; and I do
not envy the man whose heart does not some-

times quiver in unison with the bird-songs of spring. The science of ornithology is very fascinating and useful, but the unrecognized and unnamed science of bird-loving is to the more practical study what religion is to biology—the explanation of the unexplainable.

ONE day, when I was a little boy, I climbed
up the face of a rugged cliff, on a mountain-
side in North Georgia, to get some richly-
colored lichens growing there. While I was
clinging desperately to a weather-soiled projection, I chanced to see, in a small cleft near
my fingers, a gaping red-and-yellow mouth. A
chill like death swept over me and I came
near falling to certain destruction. Of course
I was well acquainted with all the snakes of
the region; what mountain-lad was not?—but
my acquaintance did not generate any desire
for familiarity with fangs and rattles, or dis-
tended heads and forked, darting tongues. A
mere glance, as my eye flashed across the
dusky little crack or fissure, carried to my
brain the impression of a wide-open, repul-
sive reptile mouth within three inches of my
bare straining fingers! nor was the glimpse,
though momentary, too slight to fix forever in
my memory a certain deadly, swaying motion
which always immediately precedes the stroke
of a venomous snake. In the course of the
merest fraction of a second I recollected a
half-dozen instances of death from the fang-
wounds of *Crotalus* or of *Toxicophis*, and an
exhaustive anticipation of the throes of disso-
lution I experienced to the full. Yet it was
not a snake, after all! So inexplicable are the
tricks of the human brain, so strange are the
sudden flashes of what one might almost dare

call intuitive knowledge, that it is not possible
to say what value should be set upon mere
impressions such as that little gaping flesh-red
and yellow mouth left indelibly burned in my
memory. Science is plodding on towards the
solution of such questions as I here raise.
With the eyes of a healthy, impressible, imaginative child I had seen a young bird gaping
over the rim of its nest, stolidly greedy for a
worm, and instantly I had grasped, without
knowing it, one of the most fascinating prob-
lems of life.

It is the fashion for scientists to pretend to
ignore the value of the imagination, and to
loudly bawl for facts; but all the knowing ones
wink under their bonnets and furtively indulge
in sublime guessing wherever the limitations
of knowledge are not set within the domain of
exactitude. Of course it would not become
me to say that a palæozoic fish cannot be described accurately with no data at hand save
the fragment of a doubtful fin-spine upon
which to build the perfect anatomy, for has
not this been done, or something very like it?
Still a rather lawless imagination can easily
enjoy the consternation with which certain
palæontological pictures might be viewed by
their draughtsmen if the original whole could
suddenly appear in the place of the precious
fossil fragment. On the other hand, however,
some of the guesses of the comparative an-
atomists may be flashes of truth revealed to
genius—that is to a simple and healthy mind.

It was years after my boyish adventure on the
cliffside that I recalled with startling vividness
its strange effect. Meantime I had been into
geology and biology and their cognate sciences,

and had studied with especial care and interest the subject of fossil birds. It now seemed
to me that my child-eyes had, in their swift
glance, seen far past that gaping young bird—
far past *Archæopteryx* and *Odontopteryx* and
Ichthyornis—to the original ancestor of the bird,
the ancient, honorable and unknown reptile.
I had received an impression of the archetype.

Sit down in the woods of spring-time and
listen to the brown-thrush or the cat-bird or,
better still, the mocking-bird, singing in the
fragrant boscage, and you may be sure that
you hear a lyre thousands upon thousands of
years old. The earth was a grand and beautiful ball of water and forests and grassy plains,
with swarms of birds and insects, and legions
of wild beasts and myriads of reptiles, a
long, dreamy, odorous and tuneful age before
man stood up in presence of his Maker and
was called good. It would be charming, if
one could but have the record of the ages all
arranged, to read the bird-songs backward (as
one may read backward through the songs of
man) to their first bubblings in the oldest
groves. Where was the first blue-bird song
uttered ? Where did the cerulean wings first
tremble among the young leaves of spring ?
It is said that science and poetry are not
friends, that they refuse to walk arm in arm,
that they scorn each other ; yet to my mind
science seems to dig up the freshest germs of
poesy, and to set free the eternal essences of
that creative force which electrifies and puts
in motion the dormant functions of genius.
Facts are dry enough and the jargon of the
doctors is not suited to enrich the poet's vo-
cabulary, but between the facts hovers the

rare, pungent, strangely powerful suggestiveness of that which fills the atmosphere surrounding facts. The chief fallacy of the scientific attitude is that which leans with confidence on the prosy for the sake of its prose,
at the same time shrinking from the poetical
on account of its poetry. The geologist feels
in some way honor-bound to avoid coming to
a picturesque conclusion with his catalogue of
facts. The catalogue must remain a catalogue.
A sense of shame would accompany any
thought of connecting imagination with his
theory of the record of the rocks.

But, despite the geologists, there is a great
deal of picturesqueness and poetry in the disclosures of the fossil beds. Set in matrices of
carbonate of lime, magnesia, silica, and the
oxides of iron, one may find the compressed
and fragmentary remains of a life that flourished before our hills and mountains were
made. This is a statement as trite, dry, and
lifeless as the fossils themselves. But when
one comes upon a mass of feathers disposed
about a strange bird-skeleton imbedded in rock
many thousands of years old, one may as well
think of what song *Archæopteryx* sang as of
what food it ate, or of how it used its long vertebrate tail. What colors had its wings and
breast and crest? Were the rectrices that
flared out on each side of the twenty vertebræ
of that strange tail dyed with rainbow hues?
These are the questions with which the scientist is ashamed to play; but the poet may ask
them of the rocks, and work out the answers,
by the rules of the imagination, to his fullest
satisfaction.

In accordance with some unchangeable law

of the scientific guild, all the beauty of our age must needs be traced back to an almost demoniac source in the palæozoic gardens of monsters, where birds had awful teeth, and where hideous saurian-like beings had wings with which to flap wildly through the poisonous air. Unfortunately enough the rocks grimly stand up, and testify for the theory of the scientists with a persistence and a lack of poetical appreciation of the beautiful truly exasperating. That there were, in those days when nature was over lusty and young, birds, fishes and reptiles fearfully and wonderfully made, cannot be for a moment doubted. It would look, to one not thoroughly learned in the records of the palæozoic ages, as if the creative power had been feeling its way, hesitating here, faltering there, gathering confidence from experience, and slowly finding out the precious secrets of life-development.

Here and there, at wide intervals, as regards both space and time, the rocks give up bird-notes, so to speak. The poet may, by holding his ear close to the strange, blurred impressions in the stones, hear the cries, the hoarse screams, the clanging trumpet-blasts of the huge land-birds and water-fowl that haunted the woods and streams and seas in that time when nearly the whole earth was a tropical region. He may hear the twitter of sparrows, too, and the careless laugh of the kingfisher.

The slab containing the remains of *Archæopteryx* is in the British Museum. It is an oblong piece of lithographic slate. The shreds of the bird lie thereon in such confusion as would mark the spot where an owl or a goshawk had eaten a blue-jay. The bones of the

head and of the sternum are not all present,
but the fragmentary wings lie in place, and one
leg with the foot attached is crooked back beside the long twenty-jointed tail. The feathers are unmistakably those of a flying bird,
and the feet are formed for tree life. It must
have been a most remarkable figure in the air,
especially if its plumage was gay-colored, with
its long, wriggling caudal streamer floating
out behind, and its claw-tipped wings spread
on either side of its reptile-like body. One
may assume that its voice was a blending of
the tones of a toad and the notes of a crow—
the first rude elements of song. Almost un-
imaginable ages have passed since the last sur-
viving *Archæopteryx* was caught in a rock matrix and forced to mould a cast for the delecta-
tion of poets and scientists. Indeed we must
refrain from attempting to span the gulf of
time between this lone relic and the next bird-
trace appearing in the earth's formations. No
more feathered vertebrate tails come to light.
Lapsing on towards the perfect form, the bird-
life, like that of certain reptiles, sloughed the
heavy caudal appendage and gathered closer
together the chief centres of its animal structure. From the cretaceous formation of the
rocks, forward to the most recent disclosures
of the caves and peat bogs, this change seems
to have gone hand in hand with a general re-
modelling of the whole sphere of mundane life.
For a vast period of time it appears that the
birds flourished, in monstrous development of
beak and teeth, the devouring demons of land
and sea. The eocene rocks furnish a wealth
of fragmentary fossils suggesting a variety of
bird-forms, mostly of giant size, waders and

swimmers as well as flyers, some of them with
jaws full of powerful teeth.   It is in this period
that nature has made indelible sketches on the
rocks, lithographic studies of her great future
work, so to speak; work that man is now so
recklessly destroying forever.   In England
the eocene has furnished a hint of the kingfishers and the heron family.   In France most
interesting discoveries have been made in the
Paris basin, and in formations of the same
horizon.   Fossil feathers, fragmentary skeletons, and even eggs, have been found, the last
mentioned in the marl deposits near Aix in
Provence.   From lacustral beds in Auvergne
and Bourbonnais a great number of birds have
come to light, nearly fifty distinct species having been described.   The marl of Ronzon has
given up an ancient plover, a gull, and a fla-
mingo, very different from presently existing
species.

Coming to our own country we step at once
amongst the choicest records of the rocks.
Beginning with the Jurassic formation, we find
in the upper beds of the period in Wyoming
the remains of a bird somewhat larger than
our well-known great blue heron (*Ardea hera-
dias*).   It was probably a toothed bird, but re-
sembled the *Ratitæ* in other respects, and was,
perhaps, not a flyer.

The cretaceous birds of America all appear
to be aquatic, and comprise some eight or a
dozen genera, and many species.   Professor
Marsh and others have found in Kansas a
large number of most interesting fossil birds,
one of them, a gigantic loon-like creature, six
feet in length from beak to toe, taken from the
yellow chalk of the Smoky-Hill river region

and from calcareous shale near Fort Wallace, is named *Hesperornis regalis.* Under the generic name *Hesperornis* have been grouped a number of species represented by skeletons more or less lacking completeness, but nearly enough perfect to show their affinities. A genus *Ichthyornis* of most remarkable toothed birds has been found in the middle cretaceous rocks of Northwestern Kansas, and a number of interesting remains have been taken from the green sand and marl beds in New Jersey. It would not serve any purpose to catalogue here all the known fossil birds. I have hastily sketched a broken outline by way of preface, leading up to what geologists call the tertiary rocks. Here we find the true ancestry of our present birds—the rocks begin to sing and twitter and chirp. Now we hear a far-away chorus, the morning voices from the old, old woods. A very breath of flowers and foliage is suggested.

In the Museum of the Boston Society of Natural History is preserved a beautiful specimen from the insect-bearing-shale of Colorado, containing a nearly complete skeleton (with feather impressions of wings and tail) of a bird belonging to the "oscine division of the *Passeres*," a division which contains all the singing birds now existing. This discovery of an oscine bird in the fossil form, dating far back of the age of man, leads the poet, not the scientist, to ask whether it may not be possible, and even probable, that some of the more ancient fossil birds had that peculiar structure of the *lower larynx,* or *syrinx,* necessary to the songster. The *oscines* are not toothed birds, and teeth have been considered an index of a low order of birds; but, on the other hand, perfectly formed wings and a well-keeled *ster-num* are the salients of the highest bird-development, and *Ichthyornis* had these, despite its teeth and fishy vertebræ. I venture to suspect that if a fairly preserved fossil skeleton, including the bill, of a poll parrot could be found in any of the mesozoic formations, no scientist would be able, without any knowledge of the parrot family save what the fossil afforded, to discover the bird's curious vocal gifts.

The perching feet of *Archæopteryx* would give it a leading characteristic of the *passeres*, and it may have had the *syrinx* of the *oscines*, despite its vertebrate, lizard-like tail; and so, too, *Ichthyornis*, notwithstanding its reptile jaws and teeth and its bi-concave *vertebræ*, may have been able to sing divinely. It was a small bird, scarcely larger than a pigeon, with a skeleton closely similar to those of the highest ornithological types, saving the *teeth* and *bi-concave vertebræ;* and who shall dispute that such a creature might have made the woods ring with its voice. True, it has been thought an aquatic bird, simply because the formation in which its remains rested is of marine origin, and on account of its teeth. There have been great changes, great progress and great retrogression, since the middle cretaceous period; but my suggestion is complete without knowing or caring about the voice of *Ichthyornis.* I have traced bird-song back into the mesozoic age, and have set the music of the rocks to ringing in the ears of my imaginative readers. If, as embryology appears to teach, the birds have come through the fish and reptile forms to their present beautiful state, by some processes of progressive evolution, the fact does not conflict with my dream. It would seem that nature has often turned back from a partly accomplished purpose, as if upon discovering a shorter and better way, and it may be that the voices of nightingale and mocking-bird have not yet reached the perfection belonging to some singer of æons ago. The syrinx of *Archæopteryx* may have been perfect, and yet the bird itself, with its cumbersome vertebrate appendage, may have been cast aside in order to begin another line of experiment, so to speak, in the direction of physical harmony. In such case the process would probably begin from the first again. It may appear that this really did take place ; for note that, after a vast geological space of time following the extermination of the highly organized *Archæopteryx*, we see the lower orders caught in the grip of the rocks, as if nature were again toiling up, but by a different route, to reach the level of the *oscines*, which appears to have been accomplished when the *Palæospiza bella* came forth in the tertiary age. This species, buried in the shale amidst the insects upon which it used to feed, may be taken as a type of the fossil song-bird and should have been named simply *Melospiza*, as the first of that genus and of the family *Fringillidæ*, just as we say, Adam or Eve !

When we come to think of it, it is next to miraculous that any traces of the palæozoic birds are left to us at all. Can we well. conceive how a sparrow or a blue-jay of our time shall be imprisoned in earth so as to be quarried out of a stone-bed some millions of years hence ? Let us pause and reflect a moment

and we shall begin to wonder how so many remains of so-called aquatic birds found their
way into the middle cretaceous beds of Kansas
and Texas.   Surely there must have been
myriads of birds in those days, else nature had
a better way then than now of taking her
dead into her bosom ?

The lower tertiary rocks of Wyoming Terri-
tory have given up an ancient woodpecker,
Uintornis lucaris, a small species, not larger
than our flicker.   He it was who drummed on
the dead trees in the lonely places of the woods
ages before the first germ that foreshadowed
man was forming under the smile of God.

Many of the ancient aquatic birds may have
built their nests in burrows, as our kingfishers
do, and various accidents may have shut them
up forever in their dens.   It can be understood how the belted halcyon of to-day might
be hermetically sealed in his burrow by the
earth falling in upon him.   Still I have heard
of but a single bone-fragment (amongst all the
fossil remains of birds) that has been referred
to the kingfisher, which argues that *Halcyon*
is a new bird in comparison with others existing at this time, or else we have not yet
chanced to cut into the banks of the old, old
brooks where he used to dig out the burrow
for his nest.

What have been called sub-fossil remains
furnish us a number of giant birds from the
sands of Madagascar and from New Zealand.
So also the peat-bogs and fens hold the bones
of rare or extinct species, principally herons
and bitterns.

Since we have been forced to study ornithology backwards, we may be said to have

just now reached the hither confine of the ancient domain of the birds. A mere outcrop here, a quarry there, with now and then a railway-cutting or a mining-drift or shaft, can afford no more than casual glimpses of what is pictured in the rocks. With *Palæospiza* as the initial, or rather the closing sketch, what if we could thumb the pages back through all the forms to *Archæopteryx* and beyond, should we not have a volume of almost weirdly unique impressions ! I have imagined that we should, in fact, find a long series of editions of the same volume, amended, remodelled, revised, but ever showing the same great development purpose. The owl was before Minerva, music was before Pan, beauty was before Venus, love was before the woman was made for Adam ; the spirit of God walked in the dawn.

The labors of A. Milne-Edwards have, to my mind, opened mines of rich suggestion to the poet as well as the philosopher and scientist, and I am sure that there is as much stimulus for the imagination as there is food for the mere reason in the discoveries of Prof. Marsh. And yet I cannot join the group who regard science as the basis of future poetry. It is not science, but the atmosphere of suggestion that stirs the pages of science, that is generative of poetry. If genius cannot see past the hard, dry fossils of to-day, far back into the living by-gone and catch those tints that are faded forever from sea and land, then genius fails at the cheapest test. It is a function of science to restore the lost head and breast bones to *Archæopteryx*, but it is the privilege of the poet to restore the colors to its feathers and to " flood its throat with song."

I have an exalted admiration of science, and place sincere trust in the outcome of its investigations; but I also sympathize most cordially with him who wishes he could have angled for Devonian fishes, or who sighs at the thought of the bird-songs of the earth's morning twilight.

But to return to our text. The curious suggestiveness of these fossil fragments of birds is not common to all the organic remains in the rocks. The cast of a delicate wing-feather in the shale of the hills, is a fertilizer of the mind and a generator of strange visions. How far that little quill has been borne down the current of time! Where was the nest with its soft lining and its wonder of green or blue or marbled eggs? Did the fragrant leaves droop over and the May-wind breathe around? Was there a brook hard by with its painted pebbles and its liquid music? Why was there no sun-burnt boy—no bare-foot girl—no cabin on the hill? I know a sportsman or two whom it would delight to shoot over a middle cretaceous marsh or shore-meadow where a good bag of *Apatornis* and *Ichthyornis* might be had! What a picnic it would be if one could prepare an ample luncheon and invite professors Gray, Coulter, Lesquereux, and many others to meet one in a jungle of the great Western Coal Basin before it was submerged! What botanizing there would be! As for me, I should like to tramp with Dr. Elliott Coues in the haunts of *Archæopteryx!* Let him collect skins while I make sketches; let him dissect fresh subjects while I listen to the voices of the strange wilderness. I should like to see the pollen of earth's first flowers upon my shoes, and hear the runic notes that have ripened into the song of the mocking-bird and the brown-thrush.

Below the surface of Professor Huxley's comparisons of the Birds and the Reptiles there is a strong current of most fascinating poetry flowing back over the fossil-bearing rocks. I take it that the first men were much nearer to Nature than we are. It may be that an hereditary far-fetched memory (so to speak) of winged monsters, suggested the dragons and griffins of early song. The crude but perfectly natural imaginings of the savages of to-day, as well as the refined fantasies of the ancients, seem to smack of this lingering hered-itament transmitted through a thousand changes from the lower estate. Pan, the goat-footed musician, is scarcely less monstrous, when we view him soberly, than many of the beings shut up in the stones.

Mr. Seeley has described a most interesting bird of the eocene period, named *Odontopteryx toliapicus*, probably a fish-eater, having nearly the habits of a cormorant, whose mouth was rimmed with bony teeth set in the powerful jaws. An expression of savage fierceness and voracity has clung to this bird's head-bones through countless ages of change. Not even the relentless grip of the rocks for a million of years could entirely quench the demoniac spirit of the creature. In what sea or lake or stream did it strike its prey? On what windy ocean crag did it rear its clamorous brood? I should like to have a look at its nest, if only to compare it with those of the fish-eaters of to-day, but much better should I enjoy a sail on the waters it haunted, with the wind on my cheek and the sharp fragrance of the salt marshes in my nostrils.

Some say that the poetry of the future will be the songs of science, that we are now in the state of transition from romance to the real. So be it if it must; but after all I should rather sing with my face to the front, if I were a poet. Science is noble and good, but the progress of the soul is better. Genius is a bird of morning, and its song is always the exponent of the most recent pulse of human passion, human knowledge of beauty, human sympathy with the joys and sorrows of the world. The rocks may give up the last secret of their hearts; the sea, too, may disgorge its treasures; but at last it is the soul of man that is the poet's field of study—the soul that walked with God upon chaos in the dark hour before the dawn of creation, the soul that still walks with him as the morning twilight slowly broadens into perfect day. It is this soul that longs backward and longs forward for the unknown, haunted all the time with some dreamy memory of its ancient chrysalis state, and feeling all the time how close it is approaching to the hour when its wings shall be full-grown.

Much has been spoken and written to demonstrate that the revelation of the rocks is or is not in conflict with the revelations of the Bible. To me the whole discussion has the ring of blasphemy. Let science go on enlightening our minds and let Christianity go on making glorious the paths of men. There is room and great need for both. Walking between the two, with a hand on the shoulder of either, let poesy gather the bird-songs and perfumes

of all the woods and fields from the beginning
to the end of time.

It is because colors have such priceless value
in the composition of the beauty our souls
crave, and because music, such as the birds
make in the dewy woods of May, goes so far
towards filling the human heart with happiness, that I close this paper with the questions:

What colors had the plumage of *Archæopteryx* ?

What song did *Palæospiza* sing?

THE END.